Robotik und Recht

edited by

Prof. Dr. Dr. Eric Hilgendorf, Universität Würzburg
Prof. Dr. Susanne Beck, LL.M., Universität Hannover

Volume 14

Eric Hilgendorf/Uwe Seidel (eds.)

Robotics, Autonomics, and the Law

Legal issues arising from the AUTONOMICS for Industry 4.0 Technology Programme of the German Federal Ministry for Economic Affairs and Energy

Nomos

The Deutsche Nationalbibliothek lists this publication in the Deutsche Nationalbibliografie; detailed bibliographic data are available on the Internet at http://dnb.d-nb.de

ISBN 978-3-8487-4196-0 (Print)
978-3-8452-8465-1 (ePDF)

British Library Cataloguing-in-Publication Data
A catalogue record for this book is available from the British Library.

ISBN 978-3-8487-4196-0 (Print)
978-3-8452-8465-1 (ePDF)

Library of Congress Cataloging-in-Publication Data
Hilgendorf , Eric / Seidel, Uwe
Robotics, Autonomics, and the Law
Legal issues arising from the AUTONOMICS for Industry 4.0 Technology Programme of the German Federal Ministry for Economic Affairs and Energy
Eric Hilgendorf / Uwe Seidel (eds.)
267 p.
Includes bibliographic references and index.

ISBN 978-3-8487-4196-0 (Print)
978-3-8452-8465-1 (ePDF)

1st Edition 2017

Introduction

AUTONOMICS for Industry 4.0 is a technology programme initiated by the German Federal Ministry for Economic Affairs and Energy (BMWi). As part of a technology competition, 14 projects with collaborators from academia and industry have qualified for funding from the Ministry for Economic Affairs and Energy. The "AUTONOMICS for *Industry 4.0*" technology programme is designed to merge state-of-the-art information and communication technology with industrial production by exploiting the potential offered by innovation in order to accelerate the development of innovative products. The aim is to strengthen Germany's leading position as a world-class production site and as a supplier of innovative production technologies. The Ministry's technology programme hence marks an important step towards the implementation of *Industry 4.0*, the Federal Government's project for the future.

In anticipation of "AUTONOMICS for *Industry 4.0*", important spadework was carried out in 2010 under the predecessor programme entitled "AUTONOMICS – autonomous and simulation-based systems for medium-sized businesses" by the Federal Ministry for Economic Affairs and Energy. The results and findings of the R&D projects in production, logistics and assembly backed by AUTONOMICS are the perfect fit for *Industry 4.0* and will serve as a basis for the "AUTONOMICS for *Industry 4.0*" programme.

The AUTONOMICS for *Industry 4.0* concept includes a host of technical, organisational and economic innovations which have been almost impossible to assess up to now in terms of their reach and impact. Although revolutions like these affect legislation, resulting in more or less quick change, the legal framework is extremely important for the development and implementation of innovation. This is particularly true in a country like Germany with its highly developed legal system based on the rule of law and very active legal policy. Laws can hinder innovation, but they can also promote it.

The legal challenges facing *Industry 4.0* are closely linked to legal issues regarding autonomous systems and, in some cases, go far beyond them. The most important difference, from a legal perspective, between the predecessor project "AUTONOMICS – autonomous and simulation-based systems for medium-sized businesses" and "AUTONOMICS for *Industry 4.0*" is the comprehensive networking that is part of the design for *Industry 4.0*. The resultant legal questions have mostly been discussed up to now in the field of Internet law, a field which has attained new and in some areas crucial importance.

Many of the legal issues brought up are closely linked to the new quality of machine networking. That is why the first step in examining the relevant legal issues understandably begins with Internet law (including provider liability law and complex cross-cutting issues involving elements of civil, criminal and public law). As in the predecessor programme, the AUTONOMICS programme throws up significant questions regarding liability for damage to machines, products and other assets as well as personal injury. From a civil law perspective, tort law and product liability law are important, however, criminal law is also relevant. It should be remembered that injured parties will often seek recourse to criminal law simply for reasons of securing evidence. In order to avoid liability (also in growing machine-to-machine communications), it is imperative that the machines and devices used within the scope of *Industry 4.0* be equipped with a host of sensors that can capture and process large quantities of ambient data and trigger suitable safety responses. The vast quantity of data captured, however, leads to considerable problems with data protection law.

With a view to liability (under civil and criminal law), the right balance between technical rules and the legal criteria for negligence must be found. Considering the high level of many technical rules, it is regrettable that up to now both worlds, i.e. technical regulation and legal regulation, have largely existed independently of each other. It is now time to transform competition into synergy. Certain sensitive data are also protected by intellectual property rights and this must also be examined. In the event of injury, applicable supplementary criminal law provisions will also become relevant.

Man-machine co-operation is extremely important within the context of *Industry 4.0*. This poses a huge challenge for labour law and especially for occupational health and safety law. All employees working within the scope of *Industry 4.0* can come into contact with smart machines. Special attention should be paid here to the considerable public relevance of the issues mentioned which can determine whether or not innovative, technically organised concepts are accepted by society. This is the reason why the relevant issues must be identified as early as possible and proposals for solutions developed. Following the first phase of analysis, these proposals can then be introduced into the societal and political debate.

The anthology „Robotics, Autonomics, and the Law“ is the result of an international workshop, held in summer 2016 in Munich, where international experts analysed the legal situation of autonomous systems and robotics in Germany, and Europe as a whole, Japan, Korea, China and the USA. The main findings are summarised in this book.

We would like to thank all participating authors for their co-operation and the Federal Ministry for Economic Affairs and Energy for their support of this project. Special thanks go to Dr. Tettenborn and Dr. Glasmacher. Moreover, we would like to thank Jochen Feldle, who helped a lot in the publication of this book, and Roger Fabry, who translated most of the texts collected.

Würzburg, 8 February 2017

Eric Hilgendorf, JMU Würzburg,
and Uwe Seidel, VDI/VDE-IT, Berlin

Table of Contents

Industry 4.0 and Law – Experiences from AUTONOMICS

Uwe Seidel,
VDI/VDE Innovation + Technik GmbH

A. Introduction

In its concentrated efforts to pave the way for the continued roll-out of *Industry 4.0*, the German Federal Government is setting a new focus in technology policy. In the AUTONOMICS for *Industry 4.0* technology programme backed by the Federal Ministry for Economic Affairs and Energy, more than 100 partners from industry and science are conducting research into new technologies and systems for industrial production. In addition, accompanying scientific research measures are addressing important cross-cutting issues related to IT security, the future world of work in *Industry 4.0*, standardisation, business models and legal frameworks. Comprehensive networking and autonomy, as part of the *Industry 4.0* concept, are posing completely new questions relating to the legal framework for interaction between smart machines, systems and workers. Data protection and IT security are just as important here as issues of liability, such as in the case of accident scenarios.

B. Industry 4.0 – New Legal Challenges in Value Chain Networks

Digitisation is a growing trend. It is rapidly transforming our industry and the services sector, creating new challenges, not only for our economy, but also for our legal system. Smart interaction between human beings and machines means that traditional production processes and business models are undergoing major transformations, as are the complex, global and often cross-sectoral value networks that are based upon them. The ideal vision of digital production involves a high level of automation, short response times and processes that run in the most effective ways possible. This 'brave new world of creating added value' offers many new opportunities for raising productivity and improving performance, but also presents a number of challenges that need to be taken into account in order to successfully implement novel types of business models.

Whenever there is a new period of innovation, legislation has difficulties keeping up with the rapid development of new technology and its possible implications; however, existing legal provisions usually suffice to limit legal risks to a reasonable degree. Both legal experts and engineers must familiarise themselves with each other's disciplines. Only if legal experts have basic knowledge of what technically controlled, autonomous processes involve can they determine in what fields new legal provisions may be required. Only if engineers have a basic level of sensitivity for detecting legal pitfalls can they turn engineered systems into successful business models. Once professionals in both of these worlds start communicating with one another, a number of grave risks can be prevented. The vague idea that there might be potential risks must not lead to innovation being slowed down. If companies were to stop optimising their production and value creation processes because they were unsure of the legal implications, this could severely impair their potential to compete on the international market. This is why the present paper also looks at how international risk assessment can be fed into existing legal provisions.

In order to be able to assess the legal impact that potential risks will have, the process scenarios and the role of individual stakeholders need to be clearly determined. Many of the reference parameters to be used for *Industry 4.0* applications still need to be defined. Traditional labour law rules continue to apply even in an age where smart machines, systems components and employees work together ever more closely. However, applying these rules in the field of innovative production – where products are constantly being enhanced – will continue to become more difficult. Machines that have been programmed in a particular way cease to be mere tools; they can make their own decisions, and select the right course of action from a wide range of different options. Smart production processes are available for use. They are controlled via sensor and actuator systems that allow for the highest possible degree of optimisation, but are insufficiently covered by current technology law. Fundamental issues such as answering the question of who will be liable for errors, material damage or physical harm need to be resolved by looking at real-life cases. After it has been determined which legal provisions are applicable, human-machine interactions can be organised in a way that ensures legal compliance. *Industry 4.0* is generating many new puzzles to be solved by the law – from civil liability, to the right to data protection, to criminal law.

In order to take account of smart factories, which are based on technically controlled, automated processes, we will need to revise our existing legal framework in a wide range of different areas. Concepts like causation and legal protection will often need to be fundamentally rethought. As in previous periods of innovation, the prevailing legislation has difficulties keeping up with the rapid development of new technology and its possible implications. The use of robots to help run production processes is taking human-machine interaction to a completely new level.

C. Supporting the Development of Compliance Rules for Technology

Industrial companies, and all persons who are in positions of responsibility within these companies, need to always strive to conduct their commercial activities in a way that is in accordance with applicable provisions and legal assessments. In order to minimise the risk of criminal or civil liability of persons who are in positions of responsibility within *Industry 4.0*, companies can set up compliance rules. First and foremost, this means compliance with the laws and agreed rules, which helps minimise a company's risk of incurring liability. The digital transformation of production and value chains creates entirely new challenges for the way smart machinery, systems and workers interact with one another. As none of these are fully covered by the existing legal framework, companies need to be as cautious as possible when assessing the risks they face.

Data protection and IT security play a key role here, as does the question of who will be liable, for example, if an accident occurs. The vague idea that there might be potential risks must not lead companies to delay innovations. If companies were to stop optimising their production and value creation processes because they are unsure of the legal implications they will have to face, this could reduce their potential to compete on the international market. Thus, non-lawyers, not least developers of technology, ought to familiarise themselves with the legal framework that applies to their work and look at how to implement their findings in order to minimise potential risks. Consequently, *Industry 4.0* requires suitable compliance rules to be put in place.

D. Examples from the AUTONOMICS Projects

Technology projects that focus on research and development often touch upon areas of technology in which no explicit solution or guidelines exist on how potential risks are to be avoided. These projects involve innovative development processes and implementation scenarios that are completely new. The AUTONOMICS for *Industry 4.0* technology programme covers a multitude of different projects and areas that reflect a wide range of technological but also legal challenges that are yet to be resolved. Companies planning to launch a new product or service need to start dealing with the legal side of things early on. This helps ensure that innovations can be turned into business models that are in accordance with the law. As has been confirmed time and again, we do not need to change the foundations of our legal system or introduce anything akin to 'legislation 4.0'. Taking into account the applicable legal framework and the laws that exist at all stages of the production process is sufficient to stay clear of any major legal pitfalls. As there still is a lack of case law on many specific issues, there is currently no clear legal framework. However, companies can interpret the existing law and apply it to their specific challenges in order to minimise risks.

In the following section, this paper will focus on a number of specific subjects in order to provide an overview of the legal challenges that have been brought to light by the projects undertaken as part of the AUTONOMICS for *Industry 4.0* technology programme.

E. Engineering Networked Production Facilities

One of the defining characteristics of *Industry 4.0* is the digital networking of production facilities. The planning, design and operation of these facilities confronts automation technicians and production engineers with completely new challenges. Flows of data – which allow for all of the development-related information and experience gained along the product life cycle to be available – are the key factor that drives the creation of value added. In order to handle the data that is obtained, new methods of dissemination and protection are needed. All of the data that is used for the optimisation of networked production facilities needs to be transmitted and processed in a legally certain manner. This means that we need to deal with data protection issues and the question of who will be liable for any damage or harm that is caused by the use of products that come out of digital production workflows.

F. Autonomous Systems in Intralogistics

Industry 4.0 also involves intralogistics – the logistics processes that take place within production facilities. The AUTONOMICS projects aim to considerably raise the level of automation of equipment used at construction sites, forklifts used on the shop floor and work pieces and machinery used in production systems. Automated intralogistics makes it possible to pinpoint the exact location of an object, identify it and prepare it for further processing. In the course of this identification process, flows of data are generated, which has legal implications.

G. Industrial Service Robotics

Service robots that can move around freely have become widely used in many areas of our everyday lives. In the context of 'industrial service robotics', a service robot is defined most often as a mobile robot that provides a service either in direct cooperation with the user or completely autonomously. In the digital factory, these robots increasingly carry out tasks that involve the controlling of production processes. Robots are becoming ever smarter (through sensors, actuators and smart systems components). They can sense when changes occur in production workflows and what needs to be changed. They can interpret events and select the right course of action from a set of different options. They can communicate with their surroundings and interact with people working in production and logistics. By using novel types of control algorithms and sensor-actuator systems to optimise machinery, production processes will be made safer and more secure, thereby reducing the number of claims that result in litigation.

H. Human Workers in Manufacturing

Although manufacturing is becoming ever more automated, *Industry 4.0* does not mean that human workers will disappear from factories altogether. As the level of automation in *Industry 4.0* increases, and interaction between humans and technology grows, employees will take on new roles and responsibilities. Rather than working side by side, humans and machines will increasingly interact with one another. As human workers become part of a system that is constantly being monitored, they will face new challenges to their labour rights. Much of the outcome of a legal assessment of potential risks or claims will depend on whether humans operating ICT-controlled production technologies are considered to be mere users of this technology or to have a substantial impact on the equipment they use.

I. Areas of Law that Relate to Autonomous Systems

German law contains an almost unimaginable number of provisions that may be relevant for *Industry 4.0.* In the following sections, the most important areas of law and the most important provisions within these areas will be described, as will be a number of relevant court decisions.

J. Civil Law

There are a number of questions that need to be answered and that will determine to a large degree how (semi-) automatic systems are to be treated under civil law. We need to distinguish between liability for a wrongful act pursuant to Section 823 *et seq.* of the German Civil Code (tortious liability) and liability for breach of contract (contractual liability). In either case, the person who is liable will be 'held to account' for the damage, provided it can be proven that the person was "at fault".

At fault (in German law), which is required for liability to be incurred, means that a person wilfully or negligently causes injury to another. A person acts wilfully if he acts deliberately, knowing or intending his action will cause damage. This type of conduct is typically expected to exist within a criminal context. As far as 'regular' business activities are concerned, it is more likely for a company to be found liable for causing damage due to negligence. A person acts negligently if he fails to fulfil his reasonable care and skills obligations, cf. Section 276 of the German Civil Code.

There are currently few statutory provisions that clearly address the issue of duty of care. *Industry 4.0* still needs to research this in the case law and relevant legal treatises and to discuss its implications. As a result, there may be situations where we face legal questions to which we can give no clear answer. In these situations, manufacturers and developers need to inform themselves about potential risks and situations where they may face liability, and to document these in order to be able to prove, in the event of a dispute, that they have adequately taken into consideration the state of the art of their technology, potential dangers and preventive measures, and that they have done so prior to any claims being brought.

As *Industry 4.0* is an area that has not yet been regulated, where (almost) no case law has been developed by the courts, the level of care to be used needs to be defined by weighing competing interests. This creates a situation of great uncertainty, as companies are faced with the possibility that the courts may later take different views on important issues. This has to do with the fact that dangerous situations can be more easily detected once the damage has been done.

Particularly in technology law, conduct that may have been considered to be acceptable at one point in time may later be considered as negligent, for example, when technological progress provides companies with completely new opportunities to prevent risks or gives rise to new types of risks previously unknown to them. As technology continues to rapidly advance, it may become even more difficult to lay down the right level of care that is required. And in the area of new technologies, the threats that companies may have to face are completely unforeseeable as the extent to which these technologies have been used in practice is rather limited.

We also need to consider the fact that the level of duty of care that is required also depends on the extent to which the users for whom a particular product is intended are familiar with that product. For example, if an autonomous system is operated only by skilled staff, a lower level of care will be required.

The civil law of negligence is not designed for dynamic, learning devices. We will still need to figure out in what cases the users, producers or even developers of a device will be held accountable for the 'actions' their devices carry out. Until there is more case law on liability, producers can only protect themselves by documenting all of their development and manufacturing processes, which will allow them to prove that they have examined all of the risk scenarios that can reasonably be expected. In this context, the question as to what monitoring obligations need to be fulfilled – for example the installation of a black box – will also need to be addressed. In addition to this, there are also a number of product liability-related issues (Act on Liability for Defective Products). For example, a product is typically defined as a tangible thing. This raises the question of whether intangible things can be considered products within the meaning of product liability rules. We also need to discuss whether the definition of the term 'defect' needs to be narrowed for software and whether we are entitled to expect a piece of software to meet the same requirements as other products. We also need to examine whether the types of defects set out in both product and producer liability rules can be applied to automated systems.

And there are still more civil law-related questions that need to be addressed: these include the controversial question of whether software can be considered a physical thing (and not only a carrier medium). Depending on how this question is answered, software may be subject to different warranty rules. And this question becomes all the more relevant as software is increasingly distributed online, for example via the Cloud. In autonomics, devices can learn by themselves and software is intended to evolve. Therefore we will need to discuss how the term 'defect' should be defined in this context.

Contractual liability means that a person will be 'held to account' for damage resulting from a breach of contract. What is particularly important about contractual liability is that if a person breaches his (contractual) obligations, the person who has suffered the damage has a right to seek compensation for the damage caused (Section 280 (1) German Civil Code). Nevertheless, the person who has caused injury does not have to pay compensation if he is not at fault (responsible for the breach of obligations that has occurred). Thus where it cannot be proved that he is at fault, having acted in a wilful or negligent manner, he escapes liability. There are only a limited number of relevant cases where liability is not fault-based in this way.

Under German civil law, there are currently no specific provisions governing the conclusion of contracts by (highly) autonomous devices or software agents. This is a problem that is already relevant in the field of financial transactions and it is very likely that it will become even more relevant as the level of automation rises (for example spare parts ordered by autonomous machinery). In order to solve this problem, new statute-making will be needed.

K. General Provisions of Civil Law

I. Section 437 German Civil Code

Section 437 of the German Civil Code is the provision that governs the rights of persons who buy defective products. It lists the rights of the buyer in case of defects: the buyer may demand rectification, revoke the agreement, reduce the purchase price, and demand compensation for damages or reimbursement of expenses. In order for a buyer to bring claims, there needs to be a contract of sale in place (Section 433 German Civil Code) and a breach needs to have occurred (Sections 434 and 435 German Civil Code). A product does not conform to the contract if it does not provide the features that one is entitled to expect.

II. Section 280 German Civil Code

Section 280 of the German Civil Code is the most important legal basis upon which claims for compensation can be made when a person breaches his duties resulting from an obligation. These can be either contractual obligations (e.g. a sales contract) or statutory care obligations (e.g. a claim for compensation in negligence). If a party breaches a duty arising from an obligation, and if the party is found to be responsible for the breach (i.e. if fault is proved), the party in breach must compensate the aggrieved party for the damage suffered. Section 280 (1) sentence 1 establishes an (objective) breach of duty as the key concept for the rights of persons to demand damages in lieu of performance. Section 280 (1) sentence 2 sets out that liability under civil law must be fault-based by referring to Section 276 of the German Civil Code. This means that a party is only responsible for the damage he has caused if he is found to have breached his obligations.

III. Section 823 German Civil Code

Section 823 of the German Civil Code is the most important legal basis upon which claims for compensation can be made in tort law. Section 823 (1) of the German Civil Code focuses on the fact that a particular legal interest has been violated (e.g. life or property). If this has been done wilfully or negligently, the guilty party will be held liable for the damage caused. In contrast to this, Section 823 (2) of the German Civil Code focuses on breach of a statutory duty intended to protect another person. The provision of Section 823 (1) of the German Civil Code states that liability is limited in that a party can only be held liable for injuring the life, body, health, freedom, property or another right of another party and for committing or failing to commit a particular act.

An act is 'unlawful' if it is contrary to the law. An act is not 'unlawful' if there are reasons that justify the act, for example if an act is committed by one party and acquiesced in by the other.

To demonstrate fault (acting wilfully = knowing and intending that an act will cause damage, negligence = failing to meet one's duty of care obligations, i.e. not being careful enough), it must be proved that a particular party has caused a particular result. This provision is not only applicable to situations where a third party suffers damage at the hands of a user of an autonomous device, but also to situations where a user or a third party suffers damage at the hands of a producer (producer liability).

IV. Section 1 Product Liability Act

Pursuant to Section 1 of the Product Liability Act, a producer of a product will be held liable if the defectiveness of his products causes a person's death, bodily harm, damage to health, or damage to an item of property. This German law, which implements Council Directive 85/374/EEC, is designed to prevent distortions of competition that arise due to diverging liability rules that exist across the various European Member States. The law is intended to promote the free movement of goods at European level. In addition to this, the law is meant to ensure that consumers will be better protected from defective products than in the past.

Under the Product Liability Act, companies may be held liable not only by consumers, but also by other companies. Although there may be cases where a party is found liable under both Section 823 of the German Civil Code and Section 1 of the Product Liability Act, the provisions on liability set out in the Product Liability Act do not go as far as those provided for in Section 823 of the German Civil Code.

V. Section 3 Product Liability Act

Section 3 of the Product Liability Act picks up upon the term 'defect' used in Section 1 of the Act, and provides a clear definition of this term. The definition uses similar criteria for defining producer liability as are used in Section 823 of the German Civil Code: producers will be held liable for defects in production, development and design, for failing to provide adequate information about the use of their products (e.g. defective guidelines, instructions, manuals etc.), as well as for products that are found to be ineffective.

In order to determine whether a product is defective, it will be taken into account what an average user is entitled to expect from the product.

L. *Criminal Law*

Under criminal law, anyone whose acts have caused damage or harm can be held to account, provided that these acts constitute a criminal offence. The manufacture, bringing onto the market and use of products in the context of *Industry 4.0* can also result in conduct that amounts to a criminal offence, for example bodily harm or manslaughter. Often, it is difficult to determine or establish causation, i.e. to prove whether a particular action is a factual cause of the result.

In addition to this, it needs to be determined whether a person has acted 'wilfully' (with intent) or 'inadvertently' (by negligence), provided that this negligent conduct is punishable by law (e.g. Section 222, 229 German Criminal Code). This last point is particularly relevant, as this means that by meeting their documentation and monitoring obligations, companies can avoid being charged with criminal negligence. Where a company can show that, in light of the state of scientific and technical knowledge at the time it put the product into circulation, it did everything possible to prevent the defect, it will, as a general rule, not be liable under criminal law. Based on the fact that in criminal law, 'causation' is defined broadly, criminal liability may be established on the part of both developers and manufacturers of defective autonomous systems, as well as on the part of programmers and sellers.

Whether a party has committed an offence that is punishable under criminal law will be determined based largely on the standard level of care to be applied under civil law. The standard level of care to be applied under civil law will always be higher than the standard level of care to be applied under criminal law. If a person's actions do not incur liability under civil law, then these actions will (most certainly) not incur liability under criminal law.

The level of care that is required will be determined based on the conduct that a reasonable, average person would engage in, were he in the situation of the wrongdoer ('objective test'). In order to prove that a person is indeed at fault, this person also needs to have failed to meet his subjective duty of care obligations. This means that, considering his individual capabilities, skills, experience and knowledge, the person could have prevented the harm that occurred. The extent to which a harmful result is foreseeable also needs to be determined both objectively and subjectively. Objectively, a result is foreseeable if an average person could have reasonably expected it to occur; subjectively, a result is foreseeable if a particular person's individual capabilities, skills, experience and knowledge should have enabled him to foresee it.

M. Fundamental Issues Relating to Liability for Defective Products

Manufacturers can be held liable for defective products. This also applies to *Industry 4.0* – which is based on networked production. Manufacturers may be held liable both under civil law (e.g. they are liable to pay compensation for the damage caused) and criminal law (e.g. they are liable for bodily injury caused by negligence). But also other persons, such as researchers and developers or in some cases even users may be at risk of being held liable for a defective product.

These persons will be held liable provided that they have caused damage to another person (e.g. the user of the product) by committing an act or by failing to act, and provided that they have acted in a way that breaches their duty. So it needs to be determined whether a certain type of conduct is unlawful. In the following sections, we will take a closer look at liability and what this means for the group of people that we are most interested in, namely for producers.

In both case law and legal doctrine, a number of duties have been developed that play a key role for products and the risks they pose. The sections below have been structured around a number of categories that are based on the categories of duties used to determine producer liability under civil law pursuant to Section 823 of the German Civil Code.

N. Duties Regarding the Design of Products

The producer of a product needs to meet particular duty of care obligations. This starts with design. The producer must not expose any third parties to risks. However, a producer is not required to ensure the maximum level of safety possible. The level of care that is required for the production of a particular product depends on the level of safety that a consumer expects the product to have. If a customer expects a product to only have an average level of safety, the producer cannot be required to provide an optimal or even maximum level of safety.

Specific requirements for product design can often be found in technical specifications and legislation. Depending on the type of product, a number of rules need to be taken into account in its design, for example DIN or ISO standards or accident prevention requirements. These can also be used to determine the state of the art of the product. If the product does not meet these requirements, this can be an indication of a breach of duty.

As far as these rules and standards that do not amount to a legal obligation are concerned, two aspects deserve special attention: even though these rules represent only minimum standards, the interest groups that design standards can set them so that they will have a direct impact on the accountability of the producer. The development of technical standards is therefore also very much relevant for the field of law. In some areas, these technical rules and standards become outdated quite quickly. And sometimes, there will be no standards yet available at the time a product is produced. This makes it difficult for producers to assess their potential liability.

Legislation such as the Product Safety Act can also contain provisions on product design. The problem here, too, is that some of this legislation sets out requirements that are not very specific and does not cover all of the details. If no specific rules or standards exist, the state of the art of a product is the guiding factor. Producers can check whether a product is designed safely by comparing it to similar products that are available on the market.

From the design stage onwards, producers should consider what groups of persons might be exposed to risks. By using a product, customers accept that they will be exposed to a certain level of risk. If, through the customer's use of the product, a third party could also be exposed to a potential risk, the producer must exercise a greater level of care. This is because the extent to which third parties can protect themselves against a defective product is limited.

Designers of a product should also take into account that users may not use the product the way it was intended. Designing a product that leaves no room for misuse of any kind does not make economic sense as it would make the product impossible to sell. Only foreseeable misuse needs to be taken into account when determining the level of care that is required of the producer. A producer bears no liability in cases where it can be proved that a product has been misused in a way that could not have been foreseen.

O. Duties Regarding Production

A production error has occurred if a product does not provide the safety that it would normally have provided had production been error-free. Here again, the level of safety that production needs to provide depends on consumers' expectations. If a producer can be expected to monitor his production process, then quality assurance is one of the most important duties that the producer must fulfil. Production errors can be caused by human error or carelessness; they can also be caused by material fatigue in production equipment. As far as networked machinery is concerned, cyber-security plays a key role. In order to combat production errors, producers may set up a monitoring system; they may, depending on the product, carry out visual inspections, measurements, or mechanical tests. In addition to this, they must check their testing equipment.

It has to be taken into account that not every error that occurs in the production environment means that a producer has failed to meet his duty of care obligations. When determining whether a producer has acted negligently, the level of acceptable risk also needs to be factored in. We live in a high-tech society and if we do not want to give up technology altogether, it will not be possible to completely prevent all risks that may arise. This is something that criminal law needs to take into account as well. Consequently, producers will not be liable for isolated cases of defective products that could not been have prevented even though all production-related requirements had been met. So if the requirement is to carry out random checks and producers comply with this requirement, then they cannot be held criminally liable for isolated cases of defective products.

P. Duties Relating to the Use of Products (Requirement for User Information)

As a general rule, producers are required to make consumers aware of and provide adequate information about the risks of their products. This is because users need to be able to protect themselves against the risks of that product. Again, consumers' expectations are the most important factor to be taken into consideration. In addition, the level of risk that a particular product poses needs to be taken into account.

By requiring producers to provide information about their products, it is to be ensured that, first of all, users will not misuse the product or use it in a way that was not intended; and secondly that users understand the risks that a particular product has even when it is used the way it was intended. In order to define the duties that producers have with regard to providing information about their products, the responsibility of producers and sellers for their products needs to be weighed against the responsibility of consumers to protect themselves.

Q. Duties that Producers have after a Product has been brought into Circulation

Producers need to fulfil a number of duties even after the product has been put into circulation. They are required to take action in order to mitigate potential risks.

If a producer discovers that a product which he has put into circulation may pose risks, there is a wide range of measures he can employ. According to case law, these measures are not limited to issuing warnings, this being namely so in cases where issuing warnings is insufficient for helping users to assess the full extent of the risk they are exposed to and to adapt their behaviour accordingly. So, in addition to issuing a warning, producers may be required to issue specific safety instructions or even recall an entire batch of a particular product. This depends on the potential severity of the harm, the likelihood of that harm occurring, and the extent to which individual users of the product can be contacted by the producer. In addition to this, the producer needs to be given some scope to specify which of his products are actually affected, with the extent to which a producer is required to take measures based on the individual case.

If the producer fails to adopt appropriate measures to prevent potential risks, he may be liable under criminal law for failure to act. He is also required to change the production of the defective product in a way that ensures that it will no longer pose a risk.

It should also be noted that producers can choose from a wide range of different options to fulfil their duties with regard to the surveillance of their products. For example, they can actively monitor the market in order to find out whether the products they put into circulation pose potential risks; they can also be asked to analyse information that is passed on to them.

R. Data Protection Law

In addition to complying with the provisions described above, companies also need to make sure they know what legislation on data protection needs to be taken into account in cases where autonomous systems are used for handling data. There are no specific provisions governing autonomous systems. The main points set out in the relevant German legislation governing data protection are virtually identical throughout. This is because the European Data Protection Directive (95/46/EC) provides that a minimum standard of data protection be maintained in all areas that are subject to regulation. The EU has adopted a new General Data Protection Regulation, which is expected to take effect in 2018.

As a general rule, personal data may only be collected, processed or stored if the data subject has consented to it or if a law permits it. For example, this means that an employer needs to seek consent from his employees before collecting and using their personal data. This can also be done by including a provision regarding consent to the use of data in the employment contract.

In *Industry 4.0*, employers collect and store huge amounts of employee data; just think of human-machine interaction where a machine needs to be precisely adapted to the needs of the employee who operates it. In this area, there are still a large number of legal questions that need to be answered.

Autonomous Systems and the Law: Why Intelligence Matters

A European Perspective

Amedeo Santosuosso,[*]
Court of Appeal of Milan; University of Pavia;
Barbara Bottalico,[**]
University of Pavia

A. Robotics and the Law Today: What is Changing

Talking about robots and the law in Europe requires a preliminary clarification of what we mean by the word 'robot'. The reason is that the scope of the definition, and the objects (and their features) we include in it, have direct effects on the legal regulation of robotics.

Defining what is a robot, is less banal than it might appear at first glance. Traditionally, embodiment was considered an essential feature of a robot: it means a machine which physically interacts with the environment. However, things are changing (and we would also say have changed) for many reasons, and, among them, mostly because of the increasing use of artificial intelligence (AI) in robotics. In other words, when we use the word 'robot', we refer to a family of entities that is dramatically growing, and today goes from systems with poor cognitive and learning capacities to systems with extremely high cognitive and learning abilities, which are able to process huge quantities of information, make decisions, and act in ways that may exceed human capacities. This is the reason why, in this paper, we maintain that stress should shift from the robot's physical body to its intelligence. This shift requires clarification of the meaning of some concepts normally used in the field, such as action, autonomy and intelligence.

---

* Professor of Law, Science and New Technologies at the University of Pavia; President of the 1st Chamber, Court of Appeal, Milan (Italy).
Contact: a.santosuosso@unipv.it.

** Post-Doctoral Fellow at the European Center for Law, Science and New Technologies (ECLT), University of Pavia, Italy; lawyer in Milan (Italy).
Contact: barbara.bottalico@unipv-lawtech.eu.

We anticipate that intelligence is important for the law because the law deals, by definition, with the actions of responsible entities. Thus knowing the level of (even artificial) intelligence of an entity is crucial in order to establish who is legally responsible both for its actions and ultimately for any harm procured by it to members of society.

In this paper, we deal with the current European legal framework for robotics, and the need for a more comprehensive concept of robotics, which might be able to take into account the increasing importance of artificial intelligence. We start by defining our field of analysis and some new features of robots related to their learning capacities (section 1). Then we sketch out the general EU legislation applicable in the field (section 2). Transnational debate with special reference to Industry 4.0 and health robotics is then discussed (section 3). Finally, (section 4) the need for a more open attitude by the EU toward technology is stressed.

I. From Robotics to Autonomous Systems: The Meaning of a Terminological Change

When talking about cognitive abilities in robotics, some notions which are used within the robotics community should be considered as a starting point.

A *cognitive robot* is an autonomous robot that executes processes analogous to human cognitive processes. It has certain abilities: (1) reasoning (the process of modification of the knowledge base of the robot through logical manipulation of available knowledge); (2) planning (computation and selection of tasks, policies and procedures for goal-directed robot behaviour) and (3) learning (the process of modification of the knowledge base of the robot gained through interaction with the environment, including people, that may produce a persistent change in the robot's behaviour).[1]

Social robots are "embodied agents that are part of a heterogeneous group, i.e. a society of robots or humans. They are able to recognise each other and engage in social interactions, they possess histories (and perceive and interpret the world in terms of their own experience), and they

1 Santosuosso A., Boscarato C., Caroleo F., Labruto R., Leroux C. *Robots, Market and Civil Liability: A European Perspective*, 2012, doi:10.1109/ROMAN.2012.6343888.

explicitly communicate with and learn from each other".[2] Evolutionary and/or developmental methods allow the autonomous development of skills in interaction with the physical (and potentially social) environment on the basis of adaptive processes driven by the environmental conditions in which the robots operate and on the basis of a utility function designed by the programmer. Of course, this interaction requires a body through which they are able to use the opportunities that their physical environment provides them with.

Evolutionary robots follow a methodology that uses evolutionary computation to develop controllers for autonomous robots, which operate in close interaction with the environment in the absence of human intervention. The idea is inspired by the Darwinian principle of selective reproduction of the fittest. Evolutionary robots use neural networks, genetic algorithms, dynamic systems, and biomorphic engineering. Experiments are in progress, and show that the survival probability (for instance in terms of access to a source of energy) of a species is affected by the behaviour of other species.[3]

In 2012, an editorial in the *Journal of Robotics* described the current state of the art and potential future perspectives and confirmed the increasing use of AI in machines:

> "… evolving and innovative theoretical tools are providing some intelligence and robustness in robotic systems similar to what is found in natural biological species. Cognition and intelligence – the ability to learn, understand, and adapt – are the creation of nature, and they play a key role in human actions as well as in many other biological species. Humans possess some robust attributes of learning and adaptation and that is what makes them so intelligent. … We are learning from the carbon-based cognitive computer – the brain – and are now in the process of inducing perception, cognition, and intelligence in robotics machines. One of our aims is to construct a robotic vehicle that can think and operate in uncertain and unstructured driving conditions.

---

2 Billard A. & Dautenhahn K. (1997), "Grounding communication in situated, social robots", *Proceedings of TIMR 97, Towards Intelligent Mobile Robots Conference*, (Technical Report Series UMCS-97-9-1), Department of Computer Science, Manchester University.

3 Pollack J.B. (2014), "Mindless Intelligence: Reflections on the future of AI" in Vargas P.A., Di Paolo E.A., Harvey I & Husbands P. (eds.) (2014), *The Horizons of Evolutionary Robotics*, MIT Press, Cambridge (Massachusetts). See also F. Silva, M. Duarte, L. Correia, S. M. Oliveira, A. L. Christensen, "Open issues in evolutionary robotics", *Evolutionary Computation* 24(2), pp. 205–236, at http://www.mitpressjournals.org/doi/pdf/10.1162/EVCO_a_00172.

> Robots in manufacturing, mining, agriculture, space and ocean exploration, and health sciences are just a few examples of challenging applications where human attributes such as cognition and intelligence can play an important role."[4]

II. The Case for Embodiment

One of the major experts in the field, Ryan Calo, in a recent paper seems to confirm the necessity of some form of physical intervention on the environment in order to define a robot:

> "a system acts upon its environment to the extent it changes that environment directly. A technology does not act, and hence is not a robot, merely by providing information in an intelligible format. It must *be* in some way. A robot in the strongest, fullest sense of the term exists in the world as a corporeal object with the capacity to exert itself physically. But again, I am talking in terms of a continuum".[5]

It is worth noting the caution of the author who, on one hand, places himself on the safe side of tradition, confirming the requisite of embodiment as a practical and operational concept, and, at the same time, admits that physical features and processing information are different features placed within a continuum.

4 Gupta, M.M., Bukovsky, I., Homma, N., Hou, Z.,& Solo, A.M.G. (2012), "Cognitive and Neural Aspects in Robotics with Applications 2011" in *Journal of Robotics*, 2012.

5 R. Calo, "Robotics and the Lessons of Cyberlaw", *California Law Review*, Vol. 103:513 2015 p.531; Ryan Calo, A. Michael Froomkin, & Ian Kerr (eds), *Robot Law*, Edward Elgar Publishing, 2016; "She, Robot: A Conversation with Helen Greiner", *Foreign Affairs*. 17 Jan. 2017. Web. 17 Jan. 2017: "You need artificial intelligence on robots, but you can have artificial intelligence purely in a disembodied state, on, say, a computer system. My belief is that the artificial intelligence in robots is going to go down a different path, one that becomes more animal-like or human-like, whereas the stuff on a computer is not competing with human intelligence; it's an adjunct to it. I think people get very confused by the language that's used by people working on artificial intelligence. They say, "It's like a brain." But it's not. It's totally a machine; it's an algorithm. When people say things like, "AI's the worst existential threat that exists for humans," I don't see it. There's nothing that I've seen right at this moment that I'm worried about being a threat to humans".

This is an interesting starting point for some clarifications. This position seems to assume that intelligence is not physical and that the only way to act in the world is through a body as a physical entity.

III. Intelligence

Systems might be grouped according to their levels of AI on a scale, which has at the lowest level tele-operated systems, autonomous systems and cognitive systems. At higher levels these systems have additional capacities such as learning, making decisions, and adapting. This latter group is characterised by the unpredictability of system actions, i.e. actions are the result of complex interactions between reactions to external stimuli and pursuit of internal goals. This is something that is usually considered intelligent behaviour.

Regarding intelligence, and its non-physical dimension, we simply highlight that there should be no doubt about the physicality of artificial intelligence which is embodied in the mechanical and computational support of a robot (as the adjective *artificial* denotes). In our view, this is sufficient in order to recognise AI as that part of a robot's body whose main feature is its ability to process information.

It does not seem necessary to deal here with the philosophical issue of the possibility of consciousness in machines:[6] the question is today at a very early (and debated) stage,[7] even though some advances in technology

6 Is mental activity a not physical feature of the brain, something over and above its neuronal base, or it is biological and therefore a physical feature of the brain? This is the typical way John Searle puts the problem: see Searle, *Intentionality, an Essay in the Philosophy of Mind*, Cambridge University Press, Cambridge, 1983; *idem*, *Minds, Brains and Science*, Harvard University Press, Cambridge, MA., 1984; *idem*, *The Rediscovery of the Mind*, MIT Press, Cambridge, MA, 1992, and other writings. Stress on the *explanatory gap* may be found in Joseph Levin, "Materialism and Qualia: The Explanatory Gap", *Pacific Philosophical Quarterly*, 64 (October), 354-61 (1983), who coined the expression. See, more recently, J. Levine "Naturalism and Dualism" in *The Blackwell Companion to Naturalism.* published online: 5 FEB 2016 DOI: 10.1002/9781118657775.ch15. Robinson, Howard, "Dualism", in Edward N. Zalta (ed.), *The Stanford Encyclopedia of Philosophy* (Winter 2016 Edition), forthcoming at http://plato.stanford.edu/archives/win2016/entries/dualism.

7 About the possibility of consciousness in machines (and its legal implications), see A. Santosuosso, "The human rights of nonhuman artificial entities: an oxymoron?"

and science might change the facts the philosophical discussion starts from.[8]

Thus we might agree with Calo's statement about the continuum, in the sense that it is recognized that it implies that AI in robots, and robots as machines, have the same origin: they are both physical features and part of a robot's body.

IV. AI and Actions

Once it is assumed that both a robot's AI and the mechanical structure of a robot are different features of a robot's body, the next questions are

---

in *Yearbook of Science and Ethics/Jahrbuch für Wissenschaft und Ethik*, De Gruyter, 2015.

8 Some examples of recent studies. The possibility of free behaviour in embodied systems that are, with no exception and at all scales of their body, subject to physical law (it is the case of a model of an artificial agent that exhibits a primitive notion of creativity and freedom in dealing with its environment) has been shown in Briegel, H.J. "On creative machines and the physical origins of freedom", *Sci Rep.* 2012, 2:522. doi: 10.1038/srep00522. See also Briegel, H.J., De Las Cuevas, G. "Projective simulation for artificial intelligence", *Sci Rep.* 2012, 2:400. doi: 10.1038/srep00400.
For the advances in image understanding see the *Journal of Computer Vision and Image Understanding* (on computer analysis of pictorial information). It publishes papers covering all aspects of image analysis from the low-level, iconic processes of early vision to the high-level, symbolic processes of recognition and interpretation). Deep Face (Facebook's facial recognition research project) can look at two photos, and irrespective of lighting or angle, can say with 97.25% accuracy whether the photos contain the same face. Humans can perform the same task with 97.53% accuracy. The DeepFace software is underpinned by an advanced deep learning neural network. For further information see the Facebook-commissioned study by Taigman, Y., Yang, M., Ranzato, M., & Wolf, Y. Taigman, M. Yang, M. Ranzato, L. Wolf, "DeepFace: Closing the Gap to Human-Level Performance in Face Verification", *Proceedings of Computer Vision and Pattern Recognition Conference* (CVPR 2014), Columbus, 2014. A T-resilience algorithm has been introduced. This is a new algorithm that allows robots to quickly and autonomously discover compensatory behaviour in unanticipated situations, cf. Sylvain Koos, Antoine Cully, Jean-Baptiste Mouret, "Fast damage recovery in robotics with the T-resilience algorithm", in *The International Journal of Robotics Research*, vol. 32, pp. 1700-1723; first published date: October-04-2013, 10.1177/0278364913499192, Among the many sources available online see *the International Journal of Machine Consciousness* (published since 2009), *The International Journal of Robotics Research* (edited at the John M Hollerbach School of Computing, University of Utah, USA).

whether and how a robot may interact with the environment and carry out actions using its AI features, either through movement of its body, or even exclusive of that body, through the exchange of information and data.

The traditional theory of embodiment is based on the (wrong) assumption that in order to change the environment it is necessary to perform a physical act and that artificial intelligence technology does not itself act. However, since the 1950s it has been known that *we can do things with words* (see the very effective title of John Austin's lectures given at Harvard)[9] and that we can act through speech (cf. also the title of John Searle's work).[10] Thus, the simple opposition between acts and information does not work in all cases. We need to explain what the context, the premises and the intention of the speech are, in order to establish whether speech is merely the expression of opinion, or whether either potentially or in reality, it is an action.

Another really complex task is to distinguish the AI of an individual robot from the AI the robot shares with the system it belongs to. For example, consider social robots (1.1. above) and then the reality of larger sociotechnical systems, where it might be unclear which robot performs an action.

V. Autonomy

In legal terms, autonomy and robots' cognitive abilities are very important particularly (but not exclusively) in order to know to what extent human control is in the loop. A recent study by Giovanni Sartor and Andrea Omicini has finally clarified some fundamental aspects of the relationship between autonomy and cognitive abilities.[11]

In the current literature, we can find various concepts of autonomy. Some focus on the ability of a machine to accomplish tasks by itself, without external direction (if autonomy is intended as the capability that en-

9 J. Austin, *How to Do Things with Words*, Oxford, Clarendon Press, 1962.

10 J. Searle, *Speech Acts*, Cambridge University Press 1969.

11 Sartor, G. and Omicini, A. (2016). "The autonomy of technological systems and responsibilities for their use" in Bhuta, N., Beck, S., Geiss, R., Kress, C., & Liu, H. Y., (eds.), *Autonomous Weapons Systems: Law, Ethics, Policy*. Cambridge University Press, 2016.

ables a particular action of a system to be automatic or, within programmed boundaries, self-governing) or the capacity to operate in the real-world environment without any form of external control, once the machine is activated and at least in some areas of operation, for extended periods of time. Others focus on cognitive capacity and in particular on the capacity to obtain new knowledge, interacting with the environment. This stresses the agent's capacity to learn what it can to compensate for partial or incorrect prior knowledge. Still others focus on the internal cognitive architecture of the system: a system is autonomous to the extent that it does not merely react to external stimuli, but rather integrates such stimuli into its cognitive structures by modifying its internal states appropriately.

Sartor and Omicini suggest that autonomy has three dimensions: independence, cognitive skills, and cognitive architecture. Cognitive architecture includes teleonomy (direction towards a goal or purpose), adaptiveness (the capacity of getting inputs from the environment and changing internal states in such a way as to better respond to challenges) and teleology (the capacity of having representations of the environment and goals to achieve as well as being able to identify appropriate means). Of course, these aspects of autonomy do not necessarily coexist and converge at all times.

These distinctions have effects on legal responsibility. For instance, a landmine (an explosive device in the ground) is completely independent in performing the operation for which it is deployed, namely, exploding when a human or vehicle passes over it. Nevertheless, it possesses very limited cognitive skills. A targeting system, on the other hand, follows the target and calculates trajectories until an operator pushes the 'fire' button. Such a system is not independent with regard to the objective of hitting the target (it is up to the operator to push the botton). However, it performs highly complex discriminative functions, distinguishing many possible input states, integrating many different features of such states to determine what circumstances are most favorable for attaining the objective, and proposing the action that is most likely to produce the desired outcome under the given circumstances.

Sartor and Omicini stress that an autonomous system is normally just a component of a larger sociotechnical system and thus the critical issue is whether a human is still in the loop of decision-making.

1.6. From Robots to Autonomous Systems

In conclusion, the term "autonomous systems" can be used broadly to encompass many kinds of artificial, decision-capable agents operating without the immediate direction of human beings. Examples include physical robots and drones as well as purely digital software agents. We chose this term in order to include both physical and non-physical robots.

The idea of the necessity of embodiment as part of the definition of a robot is based on a presumption which is wrong (see above). In any case, it is noteworthy that the lesson to be learned from speech act theory is that we can do things with words, but *not* that doing things with words is always the same as doing things through (traditional) physical actions. In other words, speech act theory (which discriminates within the area of verbal expressions) does not assume the identity of physical acts and words. It shows rather the unsoundness of the old idea that words are, by definition, external and different in kind from actions.

Thinking about the legal effects of adopting such a broad definition in our field, we might say that it has the advantage of encompassing a large spectrum of technological agents. This is, in our view, really necessary if we consider some developments in present robotics such as the increased incidence of applied artificial intelligence (which, at a certain level, implies unpredictability of action), and the interlinking of systems and machines in the environment known as the Internet of Things.

Nevertheless, all this does not imply that legal regulation should be exactly the same for all entities falling within the definition of robot, as there is great disparity in the way different systems behave. For example, injuring a worker during mechanical operations on the factory floor is very different from operating a complex city transport system, which acts mainly through the exchange of information and decision making.

B. The EU Regulatory Framework

Legislation in Europe has created a multi-layered regulatory framework. Currently, in the European Union as a whole, and in each Member State, law is made through a multifaceted law-making process involving various lawmakers at various different levels.[12] We have to consider:

a) International sources (international treaties and conventions also involving non-European countries: e.g. the World Intellectual Property Organization – WIPO (and the WIPO Convention 1967), the World Trade Organization – WTO);
b) Conventions and agreements signed within the Council of Europe (i.e. the European Convention on Human Rights);
c) European Union law
   a) *Regulations* (a legislative act of the European Union which immediately becomes enforceable as law in all Member States simultaneously);
   b) *Directives* (legislative acts of the European Union, which require Member States to achieve a particular result without dictating the means of achieving that result. Unlike Regulations (which are self-executing and do not require any implementing measures), directives normally leave Member States with a certain amount of leeway as to the exact rules to be adopted;
   c) *Recommendations* and *Opinions* (without binding force);
d) Transnational rules, i.e. legal concepts and standards that flow horizontally across national borders;
e) National legislation and sources of law (including local legislation):
   a) National legislation in its strict sense;
   b) National legislation that transposes international and/or EU sources and rules (such as directives) into national law.

In the next section, we will draw a general picture of the rules that can concern robots from a European point of view.

Here we are visualising our exploration of EU law in three concentric circles (in order to make sure that existing legal standards can respond to such critical issues). We start with the *inner circle* (constituted by EU Directive 2006/42/EC, which regulates the specific sector of machinery and, according to the commonly shared opinion, can encompass the category of robots considered as mechanical artefacts. We then go to the *middle circle*

12 This paragraph is partially based on the organisation of the legislative source presented in the paper: A. Santosuosso, C. Boscarato, F. Caroleo, R. Labruto, & C. Leroux, *op. cit.* 2012.

(constituted by more general measures governing policies to protect health, public safety and consumer interests: the EU Directive 2001/95/EC, EU Decision 768/2008/EC and EU Decision 765/2008/EC, which set the rules on product safety). Finally, we will look at the *outer circle*, which encompasses rights and guarantees recognised by EU Directive 1999/44/EC on the sale of consumer goods.

I. The Robot as a Machine: Directive 2006/42/EC on Machinery

Directive 2006/42/EC has the dual aims of harmonising the health and safety requirements applicable to machinery on the basis of a high level of protection of health and safety, while at the same time ensuring the free movement of machinery within the EU market.

Article 1 sets out the scope of the directive, that is to say the products to which the provisions of the directive are applicable. There are listed seven categories to which the directive applies: (a) machinery; (b) interchangeable equipment; (c) safety components; (d) lifting accessories; (e) chains, ropes and webbing; (f) removable mechanical transmission devices; (g) partly completed machinery.

Obviously, we will focus on "machinery", the first category, which is defined in Article 2:

> " 'machinery' means:
> - an assembly, fitted with or intended to be fitted with a drive system other than directly applied human or animal effort, consisting of linked parts or components, at least one of which moves, and which are joined together for a specific application,
> - an assembly referred to in the first indent, missing only the components to connect it on site or to sources of energy and motion,
> - an assembly referred to in the first and second indents, ready to be installed and able to function as it stands only if mounted on a means of transport, or installed in a building or a structure,
> - assemblies of machinery referred to in the first, second and third indents or partly completed machinery referred to in point (g) which, in order to achieve the same end, are arranged and controlled so that they function as an integral whole,
> - an assembly of linked parts or components, at least one of which moves and which are joined together, intended for lifting loads and whose only power source is directly applied human effort".

What seems to be essential in order to be included in this category is to be a product with parts or components linked together in an assembly. As we can see, robots, as described above, easily fit this definition.

Article 5 provides a summary of the obligations to be fulfilled by manufacturers of machinery or their authorised representatives. According to this article, before placing their products on the market or putting them into service, they have to ensure that each product satisfies the relevant essential health and safety requirements, provide the necessary information, such as instructions, and carry out the appropriate procedures for assessing conformity in accordance with Article 12. Only after the procedure for the assessment of conformity has been carried out, can the EC declaration of conformity with the directive be drafted and the CE marking affixed in accordance with Article 16.

Regulation (EC) 765/2008 defines "CE marking" as a marking by which the manufacturer indicates that the product is in conformity with the applicable requirements set out in EU harmonisation legislation providing for it.

By affixing or having affixed the CE marking, the manufacturer indicates that he takes responsibility for the conformity of the product. The CE marking consists of the initials "CE" with the graphic form shown in Annex II of the regulation.

The CE marking shall be the only marking which attests to the conformity of the product with the applicable requirements of relevant EU harmonisation legislation providing for it.

Commission communication of 20 July 2011 on the implementation of the Machinery Directive sets out the names and references of harmonised standards. In particular, the communication states the application of standard EN ISO 10218-1:2008 to robots in industrial environments (Safety requirements – Part 1: Robot – ISO 10218-1:2006, including Cor 1:2007). ISO 10218-2: 2011 (Part 2: Robot systems and integration) specifies the safety requirements for robot integration. In particular, this international standard defines the "safeguarded space" and the "restricted space" established to limit the range of motion of a machine to protect people from the hazards presented by a robotic system. Limiting devices (integral to the robot or external to it) can be mechanical or non-mechanical. But in the case of non-mechanical devices, they need to have high performance levels.

The Machinery Directive's large scope allows robots to easily be included in the categories *machinery* (Article 1, letter a) or *partly completed machinery* (Article 1, letter g). Whoever manufactures robots or places them onto the market has to comply with the procedures for assessing conformity. A manufacturer has to satisfy all relevant essential requirements and conditions set out by the dircctive (declaration of conformity, CE marking, instructions, technical file).

For the use of robots in health care, there are a specific and relevant norm and a draft ISO:

(1) Council Directive 93/42/EEC of 14 June 1993 concerning Medical Devices,[13] as amended in 2007.

Article 1 is particularly interesting for evaluating the scope of the directive as applied to autonomous systems. Medical device is indeed defined as "any instrument, apparatus, appliance, software, material or other article, whether used alone or in combination, including the software intended by its manufacturer to be used specifically for diagnostic and/or therapeutic purposes and necessary for its proper application, intended by the manufacturer to be used for human beings for the purpose of:

i diagnosis, prevention, monitoring, treatment or alleviation of disease, diagnosis, monitoring, treatment, alleviation of or compensation for an injury or handicap,
ii investigation, replacement or modification of the anatomy or of a physiological process,
iii control of conception, and which does not achieve its principal intended action in or on the human body by pharmacological, immunological or metabolic means, but which may be assisted in its function by such means"

In the healthcare context, the combination between the classification of robots as product and / or machines and the definition of medical device, open the way to the possibility of new "m-health", or medical applications of mobile devices within the scope of Directive 93/42. The norm indeed mentions "software" among the list of possible devices, and also "monitoring" and "prevention" among potential purposes.

13 http://eur-lex.europa.eu/legal-content/EN/TXT/HTML/?uri=URISERV:l21010b&from=IT.

This means that the directive could regulate m-health applications integrated into mobile devices, which would then have consequences in terms of manufacturers' and distributors' responsibilities.

In this sense, the directive lists as essential requirements: "Medical devices must meet the essential requirements contained in Annex 1 of the Directive, i.e. the necessary measures to ensure a high level of safety and performance of these devices. They must not compromise the clinical condition or the safety of patients, or the health and safety of users, or, where applicable, other persons, when they are used under the conditions and for the purposes intended. They must also achieve the performances intended by the manufacturer. In addition to the general requirements set out above, these essential requirements relate to the design and manufacture of devices, including the information which must be provided by the manufacturer to allow their safe use. These instructions can now, in certain cases and under certain conditions, be provided in electronic form."

(2) ISO Draft International Standard (DIS) 13482:
2014 – *Robots and robotic devices. Safety requirements for personal care robots*, which is intended to be a harmonised European standard related to safety requirements for robots in personal care applications. This standard would be an extension of ISO 12100, which is the general standard for the safety of machinery. ISO (DIS) 13482 defines "personal care robot" as a robot that performs aiding actions and actions contributing directly towards improvement in the quality of life of humans, excluding medical applications. It divides personal care robots in three different classes based on the level of physical interaction with users: 1) mobile servant robot (capable of travelling to perform serving tasks in interaction with humans); 2) physical assistant robot (that assists a person to perform required tasks by providing supplementation or augmentation of personal capabilities); 3) person carrier robot (has the purpose of transporting humans to an intended destination). This draft international standard covers human-robot physical contact applications, describes hazards associated with the use of these robots and provides requirements to eliminate, or adequately reduce, the risks associated with these hazards.

II. The Robot as a Product: Directive 2001/95/EC, Decision 768/2008/EC and Regulation (EC) 765/2008 on General Product Safety and Marketing of Products

Directive 2001/95/EC imposes a general safety requirement on any product put on the market for consumers or likely to be used by them, including all products that provide a service. A safe product is one which poses no threat or only a reduced threat in accordance with the nature of its use and which is acceptable in view of maintaining a high level of protection for the health and safety of persons. Once it is concluded that the Machinery Directive applies to robots, it becomes clear that robots, as machinery, are covered by the general rules fixed for the safety and marketing of products by Directive 2001/95/EC, Decision 768/2008/EC and Regulation 765/2008/EC.

A product is deemed safe when it conforms to the safety provisions provided in European legislation or, in the absence of such rules, if it complies with the specific national regulations of the Member State in which it is being marketed or sold. The product is also deemed safe if it complies with the European standard established in accordance with the procedures contained in the directive. Alternatively, the product's compliance is determined according to the following (Article 3):

a) voluntary national standards (transposing other relevant European standards), European Commission recommendations (setting out guidelines on the assessment of product safety);
b) Member States' standards in which the product is being marketed or sold;
c) codes of good practice as regards health and safety;
d) the current state of the art;
e) consumers' safety expectations.

Manufacturers may only put products on the market that comply with the general safety requirements. In addition, they must provide consumers with the necessary information in order to assess a product's inherent risks, particularly when the risks are not directly obvious (Article 5), and take necessary measures to avoid such risks (e.g. withdrawal of products from the market, informing consumers, recalling products which have already been supplied to consumers, etc.).

Distributors are also obliged to provide necessary documentation ensuring that products can be traced.

If manufacturers or distributors discover that a product is dangerous, they must notify the competent authorities and, if necessary, cooperate with them.

Member States have the power to ensure that manufacturers and distributors comply with their respective obligations (Article 6). They are to put in place structures that are responsible for monitoring product compliance with safety requirements and take necessary measures as regards product risks (e.g. prohibiting such products from being marketed). Member States set out rules to punish offenders and ensure that consumers benefit from a system that investigates complaints.

Decision 768/2008/EC also provides a range of conformity assessment procedures in the form of reference provisions, definitions and general obligations for commercial operators, from which the respective national legislatures can select the most appropriate. It also lays down rules for CE markings. This decision establishes clear definitions of fundamental concepts such as "manufacturer", "distributor", "importer", "harmonised standard", "placing on the market" and "conformity assessment". The establishment of explicit definitions applicable throughout the EU will make it easier to interpret and correctly apply future laws in this area.

In order to be placed on the market, a product must comply with certain essential requirements. The manufacturer must ensure that his products comply with the applicable requirements by carrying out or commissioning a product conformity assessment procedure. The procedures that are to be used shall be chosen from among the modules set out and specified in Annex II of the decision.

If the product complies with the essential requirements, the manufacturer affixes the CE marking to the product and draws up an EC declaration of conformity. The manufacturer indicates his name (registered trade name or registered trade mark) and his address on the product.

The product must be accompanied by instructions and safety information in language which can be easily understood.

Importers and distributors must ensure that the manufacturer has fulfilled his obligations (e.g. check that the product has a conformity marking and that the required documents have been supplied).

Manufacturers (or their authorised representatives), distributors and importers must provide the competent authorities with all necessary information on the product concerned in order to ensure product traceability.

Regulation 765/2008/EC, complementary to Decision 768/2008/EC, lays down rules on the organisation and operation of accreditation in the

Member States as well as rules on how conformity assessment bodies perform assessments of any substance, preparation or other product, transformed or not, to be placed on the EU market.

III. The Robot and the Consumer: Directive 1999/44/EC on the Sale of Consumer Goods

The purpose of this directive is the approximation of the laws, regulations and administrative provisions of Member States on certain aspects of the sale of consumer goods, such as robots, and associated guarantees in order to ensure a uniform minimum level of consumer protection in the context of the internal market.

Consumer goods must be in conformity with their contract of sale. Under article 2 goods are deemed to be in conformity with the contract if, at the moment of delivery to the consumer, they:

> "(a) comply with the description given by the seller and possess the qualities of the goods which the seller has held out to the consumer as a sample or model;
> (b) are fit for any particular purpose for which the consumer requires them and which he made known to the seller at the time of conclusion of the contract and which the seller has accepted;
> (c) are fit for the purposes for which goods of the same type are normally used;
> (d) show the quality and performance which are normal in goods of the same type and which the consumer can reasonably expect, given the nature of the goods and taking into account any public statements on the specific characteristics of the goods made about them by the seller, the producer or his representative, particularly in advertising or on labelling."

The seller shall be liable to the consumer for any lack of conformity which exists at the time the goods were delivered. In the case of a lack of conformity, the consumer shall be entitled to have the goods brought into conformity free of charge by repair or replacement or to have an appropriate reduction made in the price or shall be able to rescind the contract of sale of those goods.

Any commercial guarantee offered by a seller or producer is legally binding upon him under the conditions laid down in the guarantee document and associated advertising (Article 6). The guarantee must state that the consumer also has statutory rights and clearly state that these rights are not affected by the guarantee. Furthermore, the guarantee must state its content in simple and understandable terms, and indicate the conditions for claiming under it, notably its duration and territorial scope and the name and address of the guarantor.

C. *Fields of Application of Autonomous Systems*

In addition to the EU framework of directives and regulations, a transnational debate is in progress and involves several European countries working together on projects focusing on specific fields of application. In this section, we draw a picture of the most important documents produced, and of some important fields of application, such as Industry 4.0 and health care robots.

I. The Transnational Debate in Europe

Since 2010, a dialogue between the legal, ethical and robotics communities has taken place with increased intensity within the European Union. For instance, in 2012, the participants in euRobotics Coordination Action released the document *Suggestion for a green paper on legal issues in robotics*,[14] with the aim of providing an initial description of legal issues in robotics, together with a set of recommendations and a roadmap for future research. Critical areas were identified by the document according to the following classification: (i) robots' abilities; (ii) robots market law and consumer law (iii) intellectual property rights related to the developments of robotics in Europe (iv) labour law and robotics, (v) data privacy law and robots, (vi) criminal law, Europe and robots, (vii) conflicts and litigation involving robots (viii) non-human agents and electronic personhood.

Furthermore, in 2014 the conclusions of the European project RoboLaw[15] (funded within the 7th Framework Programme) focused on legal issues in robotics with the aim of understanding the legal and ethical implications of emerging robotic technologies.

---

14 C Leroux, R Labruto, euRobotics CA, *A Green Paper on Legal Issues in Robotics*, Grant Agreement Number: 248552, public report, Dec 2012.

15 Palmerini et al. (2014) *Guidelines on regulating robotics*. Deliverable D6.2. The RoboLaw Project, FP7, SiS 2012–2014 (GA289092). http://www.robolaw.eu/.

On 30 June 2016, the document *ELS issues in robotics and steps to consider them: Robotics and Regulations*, was released by the participants in the RockEU Robotics Coordination Action for Europe. It provided an update taking account of the fast pace of development in autonomous systems that had taken place over the previous few years.[16] The document assumed a "generative approach", aimed at finding the best way to facilitate the introduction of robots and autonomous systems onto the market. Nevertheless, we also wanted to provide tools to prevent the risks arising from autonomous systems.

The document focused on a number of selected topics, which corresponded to the main fields of application for autonomous systems today:

- robots in industry;
- robots in healthcare;
- urban robotics and self-driving cars;
- drones and military robots.

In this section, we deal with the first two topics on the above-mentioned list with the aims of: (1) briefly describing the relationship between the latest technological advancements and the law, underlining the most recent issues under debate; (2) highlighting why "intelligence matters", i.e. how the autonomy of systems can make a difference.

II. Industry 4.0 and Innovation in the Workplace

Since 2011, the term Industry 4.0 has entered the socio-economic lexicon at the European level. The idea initially arose in Germany, where government and enterprise have been investing in a new production model to strengthen and revitalise domestic manufacturing, trying also to support "re-shoring," namely encouraging the return of production to the country that had delocalised abroad in previous years.[17] During the same period,

16 B. Bottalico, A. Santosuosso, O. Goodenough, R. de Bruin, C. Holder, C. Gôme, Y. de France, C. Niel-Aubin, N. Bender, & C. Leroux, *ELS issues in robotics and steps to consider them Part 2: Robotics and Regulations*, available at https://eu-robotics.net/cms/upload/downloads/Rockeu1/2016-08-16_RockEU_deliverable_D3.4.1-part2.pdf.

17 Recommendations for implementing the strategic initiative *Industrie 4.0* are available at http://www.acatech.de/fileadmin/user_upload/Baumstruk-

the US started to move in the same direction. The *Report to the President on Ensuring American Leadership in Advanced Manufacturing*, released in June 2011 by the President's Council of Advisors on Science and Technology,[18] was the first document in which Industry 4.0, although not explicitly mentioned, helped inform a model of industry where Internet and digitalisation are crucial.

In Italy, the very first project in this direction was *Fabbrica 4.0*, which was launched in 2014 by Confindustria, the largest association of industries in Italy, with the aim of promoting better information about the possibilities that digitisation can offer to modern industries.[19] In 2016, however, Industry 4.0 became a crucial point on the agenda of the Italian government, which has promoted various initiatives and investments through the Ministry of Economic Development. In Italy, one of the difficulties in implementing Industry 4.0 has been the structure of the economy: there is a large presence of small and medium-sized enterprises and therefore a reduced capacity to invest in these new processes, which require extensive investment. Following European Union policies, the Ministry confirmed a budget of 65 million Euros to support smart projects in areas such as infrastructure/mobility and energy.[20] A project similar to Industry 4.0 is being approved by the Italian Government, and its key point is expected to be the development of enterprises in close collaboration with universities, with the aim of bringing more and more "intelligence" into factories. In the USA, the close relationship between universities and business led, for instance, to the development of *Brilliant Manufacturing*[21] software. This convinced European institutions that collaboration could only be effective if it is bidirectional.

---

tur_nach_Website/Acatech/root/de/Material_fuer_Sonderseiten/Industrie_4.0/Final_report__Industrie_4.0_accessible.pdf, (last accessed: 7.11.2016).

18 The report is available at https://www.whitehouse.gov/sites/default/files/microsites/ostp/pcast-advanced-manufacturing-june2011.pdf> (last accessed: 05/05/2015).

19 Available at http://www.confindustriasi.it/fabbrica4.0/, (last accessed: 18/11/2016).

20 A brief independent analysis of the situation in Italy at the beginning of 2016, and therefore before the implementation of the government plan, is published online by the Swiss Business Hub Italy at http://www.s-ge.com/sites/default/files/censhare_files/bericht-des-swiss-business-hub-italy-industry-situation-itali-englisch.pdf.

21 https://www.ge.com/digital/brilliant-manufacturing (last visited: 18.11.2016).

In this context, various European countries have organized parallel initiatives, such as the *Vanguard Initiative*. It is an association (initially regulated by Belgian law, it recently became an association with legal personality), which brings together the 30 most innovative regions in Europe, which alone count for two-thirds of European industry (the Italian regions participating are Lombardy and Emilia Romagna). Among its projects are 'Bio-economy'; 'Efficient and sustainable manufacturing'; 'High performance production through 3D-printing'; 'Marine renewables and offshore energy application'; 'New nano-enabled products'.[22]

A new model of production is thus emerging across Europe. The consequences of this model on work, labour and the workplace are now being analysed by jurists, economists and social scientists from an interdisciplinary perspective. In fact, the theme of high rates of automation in industry illustrates how industrial policy and employment policy cannot travel on parallel or divergent tracks, as too many aspects are mutually interconnected.

There are many disciplines that are gradually going to demand Industry 4.0 and digital manufacturing. Artificial intelligence, virtual reality, robotics, are all tools that allow the development of predictive solutions; analysis of time series data relating production, sales and product life will allow companies to change radically the way they address the market; they will gain considerable advantages from the point of view of production. Thanks to the Cloud, the interconnection between automation systems, objects and goods is getting easier and faster.

---

22 http://www.s3vanguardinitiative.eu/. The objectives were set up in the "Milan Declaration" signed in Milan in 2013. Among them: "The Vanguard Initiative brings together regions where there is a political will to play an active role in the renaissance of industry and to transform regional clusters into world-class clusters. Partner regions are committed to the development of interregional networks, based on bottom-up entrepreneurship and regional clusters that co-invest in new, interconnected European and global value chains. Our commitment to lead-by-example in inter-regional cooperation will lead to greater collaboration resulting in the co-development of mechanisms that stimulate the entrepreneurial discovery of our complementary smart specialisations, such as Smart Specialisation Platforms.".

1. New Technologies Driving Innovation

Although the last few years have been full of new inventions at very high technological levels (such as 3D printers), singling out one invention among them as an example would be neither representative nor exhaustive. The newest inventions are all linked into the so called "Internet of Things (IoT)", a computing concept that describes a situation where physical objects are connected to the Internet and are able to identify themselves to other devices and share information (Kouroupetroglou, 2014). We believe this is really a key aspect of the revolution we are describing. The significance of IoT comes from real time and significant interconnections between objects and the Internet.[23] While many of us may still think about being connected in terms of computers, tablets and smartphones, the IoT describes a world where just about anything can communicate and be connected in an intelligent fashion: the physical world becomes one big information system.

How will this affect industry and the workplace? For example, industrial machines can have sensors that notify workers when a problem occurs; surveillance systems can use locks and camera systems that inform their building managers in real time when someone is attempting to enter; plant managers can receive information and data about production continuously, and so on.

Although the Internet of Things may be considered as the main driving force of these changes in industrial organisation, attention must be drawn here to other crucial examples of innovation, starting with robotics. The expression *industrial robotics* traditionally refers to a robot arm used in a factory environment for manufacturing applications.

More recent robots and automation systems, however, are not limited by onboard computation, memory, or software. Rather than viewing robots and automated machines as isolated systems with limited computational capacity and memory,[24] *cloud robotics* suggests a new paradigm where robots exchange data and perform computation via networks. Cloud com-

---

23 An interesting definition was provided at the Lockheed Martin website: "an object that can represent itself digitally becomes something greater than the object by itself", available at http://www.lockheedmartin.co.uk/m/us/news/features/2014/isgs-internet-of-things.html (last accessed: 06/05/2015).

24 Prof. Ken Goldberg's suggestion is to consider a new paradigm where robots and automation systems exchange data and perform computation via networks, cf.

puting was indeed born as "a way of delivering computing resources as a utility service via a network, typically the Internet, which can be scaled up and down according to user requirements" (Millard 2013). Such computing resources may range from raw processing power and storage, such as servers or storage equipment, to full software applications.

Besides recent advances in connectivity among computers and robots, there is something even newer. Industrial robots are able to perform with extreme precision and speed, but they normally need to be programmed very carefully in order to complete tasks. This is difficult and time-consuming, with the consequence that robots have only been able to work in tightly controlled environments. But now there are new "intelligent" technologies coming onto the market. An article in the March 2016 issue of *MIT Technology Review* described an advanced "learning robot", released by the Japanese company Fanuc: "This kind of robot uses a technique known as deep reinforcement learning to train itself, over time, how to learn a new task. It tries picking up objects while capturing video footage of the process. Each time it succeeds or fails, it remembers how the object looked, knowledge that is used to refine a deep learning model, or a large neural network, that controls its action."[25]

The combination of the two elements previously described, i.e. "learning abilities" and connection to the Cloud, can produce very high benefits and the learning approach can be accelerated if several robots work in parallel and then share what they have learned.

Finding a shared regulatory regime across the field of robotics is not easy. This is because of the wide variety of technologies, both basic and new/advanced. It can also be difficult to separate robotics from other technologies with which it is integrated, considering that its technical components also have multiple applications. On the positive side, the collection and combination of all data generated by users and machinery with other sources and analytics can be a tremendous resource for society, policy makers, public authorities and the private market.[26] This huge potential lies in the expected predictive capacity that they bring if analysed within a

"Cloud Robotics and Automation", available at http://goldberg.berkeley.edu/cloud-robotics/ (last accessed: 07/06/2015).

25 W. Knight, "This Factory Robot Learns a New Job Overnight", *MIT Technological Review*, 16 March 2016.

26 See the document published by the Organization for Economic Cooperation and Development OECD E-Government Project GOV/PGC/EGOV(2012)7/REV1, 1

certain domain. Data are becoming more available and more adapted to the needs and capabilities of computers. Most of them are "unstructured" and Internet-based, and would not be suitable for traditional computational databases. It is said that computer tools for gathering knowledge and insights from data in this Internet era are gaining ground rapidly (Lohr, 2012).

As reported in a study conducted by the Boston Consulting Group (2015), analytics based on large data sets has emerged only recently in manufacturing, and this will have the benefit of opening up possibilities for optimising production quality, saving energy, and improving equipment service.[27] In the Industry 4.0 context, the availability of data from many different sources (production equipment and systems, enterprise and customer management systems, and so on) may become the basis for supporting real-time decision-making within production.[28]

Furthermore, big data combined with learning machines can create self-enhancing development, as predicted in 2012 by Steve Lohr in the *New York Times*: "Machine learning algorithms, for example, learn on data, and the more data, the more the machine learns" (Lohr, 2012). Learning machines are indeed particularly promising as a tool for improving the abilities of technology through the self-development of new skills, grounded on previous experience.

This very brief introduction to what technology is now offering to industry suggests taking a further step: analysing what could be the impact on the system of labour law. Is new legislation necessary to protect employees' rights? Is individuals' privacy at stake? What are going to be the effects on the job market internal and external to the workplace?

March 2013, available at http://www.oecd.org/officialdocuments/publicdisplaydocumentpdf/?cote=GOV/PGC/EGOV%282012%297/REV1&docLanguage=En (last accessed 07/06/2015).

27 The Boston Consulting Group, *Industry 4.0. The future of Productivity and Growth in Manufacturing Industries*, April 2015, available at http://www.voesi.or.at/wp-content/uploads/2013/02/Industry_40_Future_of_Productivity_April_2015_tcm80-185183.pdf (last accessed 06/06/2015).

28 *Ibid.*

2. *The Legal Challenges*

We here briefly summarise some of the main legal issues surrounding the spread of autonomous systems in industry, starting with: (1) the protection of privacy in electronic correspondence, (2) data collection and monitoring & surveillance in the workplace, (3) its impact on how work will be organised.[29]

(1) Electronic Correspondence and Privacy in the Workplace.

The problems surrounding privacy and email use have been copiously documented since the 1990s. Within the European Union, in 1989 the Committee of Ministers of Member States approved the first recommendation on the Protection of Personal Data Used for Employment Purposes,[30] but email correspondence was not yet explicitly mentioned: the recommendation was indeed mainly focused on the collection and storage/use of personal data in the recruitment process and during the working relationship, without any concrete reference to any specific technology or forms of communication.

In April 2015, the Committee of Ministers of Member States approved Recommendation CM/REC(2015)5. This starts with the assumption that REC(89)2, the previous recommendation, should be revised in light of "the increasing use of new technologies and means of electronic communication in the relations between employers and employees, and the corresponding advantages thereof...." The committee felt that the use of data processing methods by employers should be guided by principles designed to minimise any risks that such methods might pose to employees' rights and fundamental freedoms.[31]

---

29 For a wider analysis of these three issues, see: B. Bottalico "Workplace technological innovation. What role for the law?" in Santosuosso, Goodenough, Tomasi, *The Challenge in innovation in law. The Impact of Technology and science on legal studies and practise*, Pavia University Press, 2015.

30 REC (89)2 Committee of Ministers to Member States on the Protection of Personal Data used for employment purposes, available at: http://www.coe.int/t/dg3/healthbioethic/texts_and_documents/Rec%2889%292E.pdf (last accessed 2 June 2015).

31 In *Barbulescu v. Romania*, (January, 2016), the European Court for Human Rights held that a fair balance had to be struck between a worker's interest in respect for

(2) Data Collection and Monitoring & Surveillance in the Workplace.

The context of workplace monitoring is now getting extremely difficult to manage on the basis of the pre-existing legal tools. Global Positioning Systems (GPS) and Geographic Information Systems (GIS), for instance, can be used to monitor employees in activities where they are expected to travel, work outside a single office (such as in logistics and postal deliveries) or, more recently, and also to monitor the amount of work employed at specific sites, such as construction sites.

Furthermore, the combination of data concerning employees' behaviour at work with external information about them, obtained through the Internet, may affect recruitment and careers, leading to new forms of discrimination.

Within the discussion that has begun, all stakeholders seem to be aware that this time, the driving force is a new kind of innovation that is radically changing the way we live and work.[32] Thinking about the previously mentioned developments in technology, the path to follow could be that of defining the uses of the huge amounts of data that employers often have at their disposal, rather than prohibiting their collection. It is no longer useful

---

his private life and his/her mail correspondence and the contrary employer's interest in the proper functioning of the company and the activity performed by its employees. Corporate emails, phone calls and Internet used in the workplace fall within the scope of Article 8 of the European Convention on Human Rights. In this context, however, the right to privacy does not prevent the employer from checking corporate email under certain conditions, nor from giving notice to an employee of the disciplinary dismissal where it was discovered that the employee has used those means of communication during working hours, for personal purposes in violation of internal rules. Indeed, it is deemed reasonable that an employer may ensure that employees complete their assignments during working hours.

32 In Italy, for instance, up to now workers' rights have been regulated and protected by the "Workers' Statute" (Law no. 300/1970). In 2015, however, within the labour reform "Jobs Act" a law was passed to try to harmonise the existing legislation regulating controls on employees in the workplace with the possibilities opened by the new technology. The Workers' Statute was modified and employers are no longer required to obtain authorisation by the labour unions or the Labour Inspectorate for providing employees with equipment to be used to perform their work (e.g. mobile phones or personal computers), even if it is electronic and employers could use it to monitor employees' activities. However, employers are required to have a policy on the use of this kind of equipment and must comply with data protection rules.

or coherent to prohibit the collection of personal information when social networks, Internet flows, and our habits as online and offline consumers, are revealing almost all relevant details of our lives. Rather it is the use of the data that should be regulated in the sense of avoiding discrimination and misuse.

For surveillance at work, the same principle is applicable, considering how easy it has become to monitor employees' work and activities in a digitized work environment. Ensuring that data are collected anonymously where possible, and limiting their use only to legitimate reasons could be first steps toward new regulations, more open to new technologies but at the same time aimed at protecting the dignity of employees at work. The protection of the fundamental rights of workers, such as privacy and dignity, is a principle that has to be ensured in all possible scenarios. This protection of rights cannot just be ensured through the prohibition of technology that might trace physical or digital movements. It would be much better to analyse possible breaches in each workplace, according to the type of activity and the technology that is used there by employers and employees, and then delineate through specific policies how to regulate the work relationship and day-to-day activities. This could be done by providing for different regulations specifically directed to the area and technology in question, in parallel with the general provisions that have already been defined in the European context.

Indeed, the focus of Recommendation REC(2015)5 is clearly on the *purposes* rather than the intrinsic risks or capabilities of the technologies used in the workplace. This approach can be embraced since it has sufficient flexibility to be applied to various types of technology, and, at the same time, is intended to give clear directions to employers on the use of technology. Once the purposes are deemed to be legitimate, the recommendation provides for a set of "safeguards" that are supposed to be adopted in compliance with national legislation.

(3) Employability, and a New Organisation of Work

A further possible consequence of the combination of the Internet of Things with new intelligent technological developments is important changes in the organisation of industrial work. Two of the possible new developments in this area are a radical change in assembly lines and the emergence of the "smart working era" in factories (Seghezzi 2015).

In a well-known research paper released in 2013, Frey and Osborne analysed the extent to which jobs are susceptible to computerisation by implementing a methodology to estimate the probability of computerisation for 702 detailed occupations using a Gaussian process classifier. Based on these estimates, they examined expected impacts of future computerisation on US labour market outcomes. They concluded that about 47 percent of total US employment was at risk. They also provided evidence that wages and educational attainment exhibit a strong negative relationship with an occupation's probability of computerisation (Frey and Osborne 2013). Their research and its conclusions reflect the recent tendency to look suspiciously at the digitalisation of industries and the Internet of Things, and at whether they are factors of significant risk for human employment.

These scenarios can raise further considerations. A first issue, regarding skills definition, is that technological applications or machines will likely take over many human skills, making the related jobs unnecessary (it is sufficient to think of the increasing role of robots in assembly lines). In comparison with even only ten years ago, however, new forms of employment and new skills have entered the market and developed. The digital realm has entered our lives and our activities in a revolutionary way, and it requires more and more experts in the field who can provide guidance both as an external service and as an internal management tool.

The digital scenario is forcing many workers to augment their skills. On the one hand, some human skills could be devalued (data managing jobs, salespeople, bank clerks) as a consequence of increasing on-line consumption (e-commerce, e-banking). The use of machines for physical activities that do not require specific human input (e.g. industry robots) will affect manufacturing employment. On the other hand, some other professional roles are emerging in totally new environments: cloud experts, app developers, extra-specific programmers and so on. Furthermore, the skills required of a single worker are likely to evolve in the sense of including digital competences irrespective of the employee's specific role in the company.

A second point can be made about work organisation: the digitisation of work can easily lead to the exploitation of the workforce in very destabilizing ways. However, the possibility for work to be conducted on digital platforms without the need for expensive physical premises opens extraordinary opportunities for companies in developing countries, which, in turn, are increasingly hiring European companies to develop innovative IT solutions and manage services online.

Furthermore, the possibility of working from somewhere other than company offices opens up new scenarios for telecommuting and teleworking, insofar as it makes it possible to work from home some days per week or per month, helping families care for children or elderly relatives and disabled people without limiting individuals' career prospects.

3.3. Robotics and Health Care

In 2008, the European Commission published the results of a consultation with experts and stakeholders which identified 21 main innovation areas that could be considered key product/market combinations and from which six representative areas could be regarded as ripe for further investigation and road-mapping:

- Smart medical capsules (for endoscopy, biopsy and targeted drug delivery);
- Intelligent prosthetics;
- Robotised patient monitoring systems;
- Robotised surgery (a combination of the areas related to the facilitation of the surgeon in the operating room);
- Robotised motor coordination analysis and therapy;
- Robot assisted mental, cognitive and social therapy.[33]

In these instances, it is particularly interesting to use the expression "autonomous agents" because the technological innovations also concern apps, learning and autonomous devices for telemedicine and further technologies which interact with "physical robots".

The field of robotics and telemedicine is particularly challenging, from a legal and ethical point of view.

There are currently three types of possible applications in the field:

a. Devices programmed to check patients' physical condition and indicate the dosage of medicines or, more generally, what actions to carry out;

33 *EC-robotics for Healthcare Final Report*, 2008: http://www.ehealthnews.eu/images/stories/robotics-final-report.pdf.

b. Robots designed to provide remotely-located physicians with information on how their patients are recovering when treated at area hospitals;
c. "Care-robots", i.e. robots specifically designed to assist elderly or disabled people (without any strictly therapeutic purpose).[34]

As a consequence, some key legal issues arise with regard to telemedicine devices and software, such as: who is responsible if the patient does not follow the proposed treatment plan? Who is responsible if the patient follows the plan but her condition does not improve or worsens?

The most important innovation in this technological landscape is a new generation of adaptive ("intelligent") devices. These devices can learn over time, according to the data provided by the patient and by the physician. In the past, a manufacturing defect had a "stable" dimension, but now we can hypothesise the existence of a spectrum of responsibility. In particular, if the functioning of autonomous systems also depends on data input by the user, the patient could be found more liable for damage the longer the time that has passed since he started using the device. In other words, the patient's responsibility would increase over time.

4. Why Intelligence Matters in Autonomous Systems

In this paper, we have presented some basic concepts and definitions in the field of robotics and law and maintained that the umbrella concept should shift from *robot* to *autonomous system*. We have argued that this change is necessary because of factual technological innovation (the increasing rate of AI in physical robots still having a physical body) and a theoretical change in the meaning of the concept of *action*, which we assume (according to speech act theory) can be performed both physically and *with words*.

Our conclusion is that the expression "*autonomous system*" reflects a more comprehensive concept of what an action is and what the body of a robot is. Thus it can provide a conceptual and legal framework for dealing with the expansive introduction of new technologies in our societies.

---

34 The industry of care-robots is growing exponentially. According to *The Business Insider*, one-third of the Japanese government's budget is allocated to developing care-robots. D. Muoio, "Japan is running out of people to take care of the elderly, so it's making robots instead", *The Business Insider UK*, 20.11.2015.

In the second and third sections we have presented a general picture of European regulation and the main law-related issues in the field.

In our view, while debate is intense in the European Union, European Member States should take concrete steps to facilitate the introduction of these technological devices in their societies and, in doing so, take advantage of them for the development of human relationships and economic growth.

Bibliography

Austin J., *How to Do Things with Words*, Oxford, Clarendon Press, 1962.

Billard A. & Dautenhahn K. (1997), "Grounding Communication in Situated, Social Robots", *Proc. of TIMR 97, Towards Intelligent Mobile Robots Conference*, (Technical Report Series UMCS-97-9-1), Dept. of Computer Science, Manchester University.

Bottalico B. "Workplace Technological Innovation. What Role for the Law?" in Santosuosso, Goodenough, Tomasi, *The Challenge in Innovation in Law. The Impact of Technology and Science on Legal Studies and Practise*, Pavia University Press, 2015.

Bottalico B., Santosuosso A., Goodenough O., de Bruin R., Holder C., Gôme C., de France Y., Niel-Aubin C., Bender N., Leroux C., "ELS Issues in Robotics and Steps to Consider them Part 2: Robotics and Regulations", available at https://eu-robotics.net/cms/upload/downloads/Rockeu1/2016-08-16_RockEU_deliverable_D3.4.1-part2.pdf.

Briegel, H.J. "On Creative Machines and the Physical Origins of Freedom", *Sci Rep.* 2012;2:522. doi: 10.1038/srep00522.

Briegel, H.J., De Las Cuevas, G. "Projective Simulation for Artificial Intelligence", *Sci Rep.* 2012;2:400. doi: 10.1038/srep00400.

Calo R., "Robotics and the Lessons of Cyberlaw", *California Law Review*, Vol. 103:513 2015 p. 531.

Ryan Calo, A. Michael Froomkin, & Ian Kerr (eds), *Robot Law*, Edward Elgar Publishing, 2016.

Gupta, M.M., Bukovsky, I., Homma, N., Hou, Z.,& Solo, A.M.G. (2012), "Cognitive and Neural Aspects in Robotics with Applications 2011" in *Journal of Robotics*, 2012.

Harvey I & Husbands P. (eds) (2014), *The Horizons of Evolutionary Robotics*, MIT Press, Cambridge (Massachusetts).

Howard R., "Dualism", in Edward N. Zalta (ed.), *The Stanford Encyclopedia of Philosophy* (Winter 2016 Edition), forthcoming at http://plato.stanford.edu/archives/win2016/entries/dualism.

Knight W., "This Factory Robot Learns a New Job Overnight", *MIT Technological Review*, 16 March 2016.

Koos S., Cully A., Mouret J.B. (2013) "Fast Damage Recovery in Robotics with the T-Resilience Algorithm". *International Journal of Robotics Research*, 32(14): 1700-1723.

Leroux C, R Labruto, euRobotics CA, *A Green Paper on Legal Issues in Robotics*, Grant Agreement Number: 248552, public report, Dec 2012.

Levin J., "Materialism and Qualia: The Explanatory Gap", *Pacific Philosophical Quarterly*, 64 (October), 1983, pp. 354-361.

Levine J. "Naturalism and Dualism" in *The Blackwell Companion to Naturalism.* published online: 5 FEB 2016 DOI: 10.1002/9781118657775.ch15.

Muoio D., "Japan is Running Out of People to Take Care of the Elderly, so it's Making Robots Instead", *The Business Insider UK*, 20.11.2015.

Palmerini E. et al. (2014) *Guidelines on Regulating Robotics*. Deliverable D6.2. The RoboLaw Project, FP7, SiS 2012–2014 (GA289092). http://www.robolaw.eu/.

Pollack J.B. (2014), "Mindless Intelligence: Reflections on the Future of AI" in Vargas P.A.,Di Paolo E.A., Harvey I., Husband P., *The Horizons of Evolutionary Robotics*, MIT Press 2014.

Santosuosso A., "The Human Rights of Nonhuman Artificial Entities: an Oxymoron?" in *Yearbook of Science and Ethics/Jahrbuch für Wissenschaft und Ethik*, De Gruyter, 2015.

Santosuosso A.; Boscarato C.; Caroleo F.; Labruto R.; Leroux, *Robots, Market and Civil Liability: A European Perspective, C.2012, doi:10.1109/ROMAN. 2012.6343888*

Sartor, G. and Omicini, A. (2016). "The Autonomy of Technological Systems and Responsibilities for their Use" in Bhuta, N., Beck, S., Geiss, R., Kress, C., & Liu, H. Y., (eds.), *Autonomous Weapons Systems: Law, Ethics, Policy*. Cambridge University Press, 2016.

Searle J., *Intentionality, an Essay in the Philosophy of Mind*, Cambridge University Press, Cambridge, 1983;

Searle J., *Minds, Brains and Science*, Harvard University Press, Cambridge, MA., 1984;

Searle J., *Speech Acts*, Cambridge University Press 1969.

Searle J., *The Rediscovery of the Mind*, MIT Press, Cambridge, MA, 1992, and other writings.

Silva F., Miguel D., Lu´ıs C., Sancho M.O., Anders L.C, "Open Issues in Evolutionary Robotics", *Evolutionary Computation* 24(2), pp. 205–236, at http://www.mitpressjournals.org/doi/pdf/10.1162/EVCO_a_00172.

Taigman, Y., Yang, M., Ranzato, M., & Wolf,Y. Taigman, M. Yang, M. Ranzato, L. Wolf, "DeepFace: Closing the Gap to Human-Level Performance in Face Verification". *Proceedings. of Computer Vision and Pattern Recognition Conference* (CVPR 2014), Columbus, 2014.

Robots in American Law

Ryan Calo,[*]
University of Washington

This article closely examines a half century of case law involving robots—just in time for the technology itself to enter the mainstream. Most of the cases involving robots have never found their way into legal scholarship. And yet, taken collectively, these cases reveal much about the assumptions and limitations of our legal system. Robots blur the line between people and instrument, for instance, and faulty notions about robots lead jurists to questionable or contradictory results.

The article generates in all nine case studies. The first set highlights the role of robots as the objects of American law. Among other issues, courts have had to decide whether robots represent something "animate" for purposes of import tariffs, whether robots can "perform" as that term is understood in the context of a state tax on performance halls, and whether a salvage team "possesses" a shipwreck it visits with an unmanned submarine.

The second set of case studies focuses on robots as the subjects of judicial imagination. These examples explore the versatile, often pejorative role robots play in judicial reasoning itself. Judges need not be robots in court, for instance, or apply the law robotically. The robotic witness is not

---

* Assistant Professor of Law, Assistant Professor (by courtesy) of Information Science, University of Washington. Affiliate Scholar, Yale Law School Information Society Project and Stanford Law School Center for Internet and Society. Senior Fellow, New America. This paper will benefit from presentation at University of Chicago School of Law, Michigan Law School, University of Texas School of Law, and various other seminars, conferences, and workshops in 2016. I would like to thank participants in these events as well as David Brownell, Jean Brownell, Danielle Keats Citron, Woodrow Hartzog, and Amanda Levendowski for helpful edits and comments. Thank you also to Noemi Chanda, Ella Corren, and the University of Washington Gallagher Law Library for wonderful research assistance.
Originally published: Calo, Ryan, Robots in American Law (February 24, 2016). University of Washington School of Law Research Paper No. 2016-04. Available at SSRN: https://ssrn.com/abstract=2737598.

to be trusted. And people who commit crimes under the robotic control of another might avoid sanction.

Together these case studies paint a nuanced picture of the way courts think about an increasingly important technology. Themes and questions emerge that illuminate the path of robotics law and test its central claims to date. The article concludes that jurists on the whole possess poor, increasingly outdated views about robots and hence will not be well positioned to address the novel challenges they continue to pose.

INTRODUCTION

"Robots again." So begins Judge Alex Kozinski's dissent from the Ninth Circuit's decision not to rehear *Wendt v. Host International* en banc.[1] "Robots," because *Wendt* involved animatronic versions of two popular television characters that, the actors said, violated their rights of publicity.[2] "Again," because, just a few years before, the Ninth Circuit decided *White v. Samsung*, in which Vanna White sued the electronics giant Samsung for featuring a robot version of the game show host in an advertisement.[3]

Robots appear in surprising number and variety in American law. *White* and *Wendt* are fairly well known. But most of the cases to involve robots have never made their way into legal scholarship. This article closely examines six decades of courts struggling with robots—just in time for the technology itself to enter the mainstream. This detailed examination leads to a simple thesis: robots confront courts with unique legal challenges that judges are not well positioned to address.

The challenges robots pose will only become more acute in light of the explosive growth of the robotics industry over the next decade. Today robots are leaving the factory and theatre of war and entering our roads, skies, offices, and homes. We are in the midst of a robotics revolution.[4] Popular technology companies are investing billions in robotics and artifi-

---

1 Wendt v. Host Int'l, Inc., 197 F.3d 1284 (9th Cir. 1999) (Kozinski, J., dissenting from denial of petition for rehearing).

2 Wendt v. Host Int'l, Inc., 125 F.3d 806, 809 (9th Cir. 1997).

3 White v. Samsung Elec. Am, Inc., 971 F.2d 1395 (9th Cir. 1992), *cert. denied*, 508 U.S. 951 (1993).

4 *See* Gill A. Pratt, *Is a Cambrian Explosion Coming for Robotics?*, 29 J. ECON. PERSP., 51, 51 (2015) ("Today, technological developments on several fronts are fomenting [an] explosion in the diversification and applicability of robotics.").

cial intelligence.[5] Patent filings for robots are skyrocketing.[6] Hardly a day goes by without a national headline devoted to driverless cars or drones.[7]

In previous work, I examined what the mainstreaming of robotics might mean for American law and legal institutions.[8] I grounded the discussion in the legal and policy fallout from the last transformative technology of our time, the Internet. The Internet has a set of core qualities that tended to pose challenges for law. For example, the Internet allows for instant exchange of goods and services across borders, which invited courts to revisit the rules of jurisdiction.[9] Robotics, I argued, has a different set of core qualities than the Internet and, accordingly, will generate new puzzles for law.[10]

Robotics and the Lessons of Cyberlaw drew several thoughtful responses.[11] Yale Law professor Jack Balkin agrees that robots will generate interesting new legal puzzles but questions whether we can know what these might be in advance.[12] Balkin observes that the only reason we can point to the legally relevant features of the Internet is that we have two decades

5 Google, Amazon, Facebook, IBM, and many other companies are each investing millions or billions of dollars in these technologies. *See, e.g.*, John Markoff, *Google Puts Money On Robots, Using the Man Behind Android*, N.Y. TIMES, Dec. 4, 2013; John Letzing, *Amazon Adds That Robotic Touch*, WALL ST. J.., Mar. 20, 2012. Some recent investments outside of the United States are even more dramatic. *See, e.g.*, *EU launches world's largest civilian robotics programme*, EUROPEAN COMMISSION, June 3, 2014, http://europa.eu/rapid/press-release_IP-14-619_en.htm; Kelvin Chan, *China's robot revolution is happening*, ASSO. PRESS, Sep. 23, 2015.

6 *See* "World Intellectual Property Report: Breakthrough Innovation and Economic Growth," WORLD INTELLECTUAL PROPERTY ORGANIZATION, 120-35 (2015) (discussing surge in patent and other intellectual property activity).

7 A search of Westlaw revealed well over four thousand news stories where drones or driverless cars appeared in the headline in 2015 alone.

8 Ryan Calo, *Robotics and the Lessons of Cyberlaw*, 103 CALIF. L. REV. 513 (2015).

9 *Id.* at 520-21.

10 *See generally id.*

11 For example, bestselling science fiction writer Cory Doctorow responded to the article by calling into question whether there can be any legal distinction between robots and computers. Cory Doctorow, *Why it is not possible to regulate robots,* THE GUARDIAN, Apr. 2, 2014.

12 Jack M. Balkin, *The Path of Robotics Law*, 6 CALIF. L. REV. CIRCUIT 45, 49 (2015).

of experience with theorists and courts writing about cyberlaw.[13] For Balkin, we will have to wait and see what path robotics law will follow. Or perhaps look to science fiction, where the laws of robotics are familiar.[14]

Professor Balkin's helpful critique led to this article's animating question: Are we in the same place today with robotics that we found ourselves at the dawn of cyberlaw? I submit that we are not. Courts and law professors in the 1990s had very limited experience with the Internet; it was brand new.[15] Robots, in contrast, have played a role in American society since at least the 1950s. And, like most technologies, they have been involved in legal disputes. What do those disputes tell us today, as robots enter a new golden age? And how should courts and other jurists think about contemporary robotics?

We should not be surprised that American courts have had to make decisions involving robots. People have been thinking about robots for thousands of years. The sixth century manuscript *Shai Shih t'u Ching* catalogues mechanical orchestras and other automata that predate the birth of Christ.[16] Since at least the 1980s, robots have been instrumental in keeping American factories competitive.[17] In the past two decades, the United States government has embraced robotics in its effort to overhaul the most powerful military in the world.[18]

What *is* surprising is how difficult and complex these cases turn out to be, and what they wind up saying about the law itself. This article canvasses hundreds of decisions concerning robots over half a century, generating nine new case studies for the burgeoning field of robotics law. The first six consist of archetypes of a particular kind of problem that robots pose as objects. Among other issues, courts have had to decide whether

---

13 *Id.* at 48.

14 *Id.* At 60.

15 *See* Frank H. Easterbrook, *Cyberspace and the Law of the Horse*, 1996 U. CHI. LEGAL F. 207, 207 (calling attention to legal scholars' lack of experience with the Internet).

16 *See* Pau Alsina, *Arte*, CIENCIA, Y TECHNOLOGIA 85 (2007) (discussing the Shai Shih t'u Ching or "Book of Hydraulic Excellencies"). *See also* IBN AL-RAZZAZ AL-JAZARI, *The Book of Knowledge of Ingenious Mechanical Devices* (Donald R. Hill, trans. 1974).

17 [cite].

18 *See* PETER W. SINGER, WIRED FOR WAR: THE ROBOTICS REVOLUTION AND CONFLICT IN THE TWENTY-FIRST CENTURY (2009).

robots represent something "animate" for purposes of import tariffs,[19] whether robots can "perform" as that term is understood in the context of a state tax on performance halls,[20] and whether a salvage team "possesses" a shipwreck by visiting it with an unmanned submarine.[21]

Robots also play an interesting role as the subjects of judicial imagination. The article's remaining three case studies explore the versatile, often pejorative role robots play in judicial reasoning itself. Judges need not be robots in court, for instance, or apply the law robotically.[22] The robotic witness is not to be trusted.[23] People who commit crimes under the robotic control of another might avoid sanction.[24]

Together these case studies paint a nuanced picture of the way courts think about an increasingly important technology. Themes and questions emerge that illuminate the path of robotics law and test its central claims to date. We can see already how robots begin to blur the line between people and instrument, and how faulty assumptions about robots lead jurists to questionable or contradictory results. The article concludes that jurists on the whole possess poor, increasingly outdated views about robots and hence will not be well positioned to address the novel challenges they continue to pose. A more comprehensive understanding of robotics and the diversification of sources of robotics law to include legislators and regulators may help—hopefully in enough time to a make a difference.

The article proceeds as follows. Part I discusses robots as the objects of law, i.e., as things in the world that occasion legal disputes, placing particular emphasis on the role of robots as surrogates for people.[25] We begin with *White* and *Wendt* but quickly turn to less-examined legal territory.

---

19 *E.g.*, Louis Marx & Co. v. United States, 40 Cust. Ct. 610, 1958 WL 8607 (1958). *See also* infra.

20 Comptroller of the Treasury v. Family Entertainment Center of Essex, Inc., 519 A. 2d 1337 (Md. 1987). *See also infra.*

21 Columbus-America Discovery Group, Inc. v. Abandoned Vessel, S.S. Central America, 1989 A.M.C. 1955 (1989). *See also infra.*

22 *E.g.*, Commonwealth of Williams v. Local Union 542, 388 F. Supp. 155 (1974); Allen v. State, 290 Ala. 339 (1973) ("The trial judge is a human being, not an automaton or a robot."). *See also* infra.

23 *E.g.*, Rong Lin v. Mukasey, 299 F. App'x 10 (2d Cir. 2008). *See also* infra.

24 *E.g.*, Frye v. Baskin, 231 S.W. 2d 630 (Mo. App. 1950). *See also* infra.

25 Woodrow Hartzog offers this helpful term. Balkin also talks about the notion of a "substitution effect" where robots stand in for people. Balkin, *supra* note 12, at 55. In previous work, I speak in terms of the social valence of robots, i.e., the unique status of robots as an artifact that feels like an animate being. Calo, *supra* note 8,

Part II investigates the role of robots subjects in judicial reasoning, i.e., as metaphors or analogies that actually drive the decisions of courts. The concept of a robot in these cases is not a mere passing reference but a part of the central holding, an idea cited by later courts for its binding or persuasive effect. Part III unites the past of American robot law with its likely future—a crucial exercise given the rapid mainstreaming of drones, driverless cars, surgical robots, home robots, and other robotics technology. This Part examines whether past cases shed light on existing puzzles, generates new questions for scholars including around race and gender, and discusses the role of courts, legislators, regulators, and others in setting the path of robotics law going forward. A final section concludes.

A. ROBOTS AS LEGAL OBJECTS

This Part consists of six case studies generated by an analysis of over two hundred cases involving robots and analogs. More specifically, my research assistants and I searched Westlaw for opinions that mention robots and close synonyms such as "robotics" and "automaton." For the case to be a candidate for analysis, the word could not appear merely in the body of the decision but had to appear in the syllabus or headnotes. In this way, the research attempts to filter out hundreds of other cases in which the term "robot" appears but does not meaningfully factor into the factual or legal dispute.[26]

The concept of a robot is not without controversy. In my previous work, I embrace a definition of a robot as a machine with three qualities: (1) a robot can sense its environment, (2) a robot has the capacity to process the information it senses, and (3) a robot is organized to act directly upon its environment.[27] I feel this definition—to which the literature refers as the "sense, think, act" paradigm—best reflects how robots differ from previous or constituent technologies such as a laptop.[28] For the purposes of this

at 545-49. The tendency is so strong that soldiers have reportedly risked their lives on the battlefield to rescue a robotic member of the team. *Id.* at 515.

26 This is an admittedly imperfect heuristic. I acknowledge the limitations of using headnotes and syllabi, written not by the courts themselves, but by lawyers after the fact.

27 Calo, *supra* note 8, at 529-32.

28 *Id.* at 529.

article, however, I did not use my own definition to screen out any technology. Rather, I looked for the court to use the word robot or a close synonym. However, I did exclude cases—such as *CNET Networks, Inc. v. Etlize, Inc.*—where the "robot" at issue referred exclusively to a software program running in the background of a server or website.[29] The robots that follow tend to be embodied, physically, in the real world.

Even with these various heuristics in place, the role of the robot in the majority of cases is best characterized as incidental. By incidental, I mean that the case would likely have come out exactly the same way were the technology at issue not a robot but some other object or concept. Maybe it's a copyright case where robots happen to figure into the plot that the plaintiff alleges was unlawfully copied.[30] Or perhaps it is a case of medical malpractice where the surgeon operated on the plaintiff with a surgical robot.[31] The movie plot could involve aliens or the surgery happen manually and present the court with the identical legal issue.

Many of these incidental cases are quite interesting. In *Robotic Vision Systems, Inc. v. Cybo Systems, Inc.*, for instance, a client of a robotics firm sued because, rather than send human technicians to resolve an installation problem, the robotics firm sent two robots named Al Bove and Al Treu.[32] The client found the robots annoying and unhelpful and sued for breach of contract. In *Reinhardt v. Fuller*, a criminal defendant fired four shotgun blasts at a police robot during his arrest.[33] Robotic props have repeatedly caused injuries on stage and film by behaving unexpectedly, including on the set of a movie *about machines that came alive and hurt people*.[34] These are colorful facts. Nevertheless, the issues tend to turn on standard

---

29 *See* CNET Networks, Inc. v. Etlize, Inc., 547 F. Supp. 2d 1055, 1065 (N.D. Cal. 2008) (defining the term "crawler" in the context of a patent dispute).

30 *E.g.*, FASA Corp. v. Playmates Toys, Inc., 869 F. Supp. 1334 (N.D. Ill 1994) (finding that "robot-like battle machines" are familiar themes not subject to copyright protection).

31 *E.g.*, Balding v. Tarter, 3 N.E.3d 794 (Ill. 2014) (denying an appeal from a medical malpractice case involving robot-assisted prostate surgery).

32 F.Supp.2d 151 (E.D.N.Y. 1998).

33 2008 WL 5386802 (E.D. Cal. 2008).

34 "[O]n the set of a motion picture entitled Maximum Overdrive," the plaintiff "sustained severe and permanent damage to his 'shooting eye' when a remote controlled powered lawnmower lost control." Nannuzzi v. King et al., 660 F. Supp. 1445, 1446 (S.D.N.Y. 1987) (remanding action to state court). *See also* Provenzano v. Pearlman, Apat, & Futterman LLP, 2008 WL 4724581 (E.D.N.Y. 2008) (unreported legal malpractice case where plaintiff failed to recover against manu-

principles of contract, criminal, and tort law. What distinguishes the cases that follow is that they turn in some way on the unique features of robots.

In short, I used a particular heuristic to sort cases that involve robots into two categories: cases where the existence of a robot was incidental and cases where the robot was arguably instrumental. From the latter I generated the six case studies that follow.[35] In this Part, I present these cases and offer commentary on the court's analysis.

I. Appropriation by Robot

I have already mentioned what is perhaps the most visible case involving a robot, that of *White v. Samsung*.[36] In the early 1990s, the electronics giant Samsung ran an amusing series of print ads speculating about the future. One depicted a "female-shaped robot ... wearing a long gown, blonde wig, and large jewelry" standing on what appeared to be the set of the game show *Wheel of Fortune*.[37] The caption read "Longestrunning game show. 2012 A.D."[38] The host of *Wheel of Fortune*, Vanna White, sued Samsung in federal district court for violating her right of publicity and falsely implying an endorsement. The trial court rejected her claims on summary judgment and White appealed.

The Ninth Circuit reversed. The appellate court agreed with the district court that, for purposes of California's right to publicity statute, Samsung

facturer when robotic camera struck her in the head); Paul McCann, *TV robot injuries studio workers*, THE N.Y. TIMES, Jan. 8, 2000 ("During filming recently, a 170 lb robot came to life after it was switched off and careered out of control, injuring a stage technician. He needed stitches for an ankle wound caused by sharp spikes on the robot.").

35 The closest set of methodology commitments to my approach are likely grounded theory, which recommends approaching a data set without a preconceived hypothesis, and thematic network analysis. For a discussion of grounded theory, see ANSELM STRAUSS & JULIET CORBIN, GROUND THEORY METHODOLOGY: AN OVERVIEW, IN HANDBOOK OF QUALITATIVE RESEARCH (1994). For a discussion of thematic network analysis, see JENNIFER ATTRIDE-STIRLING, THEMATIC NETWORKS: AN ANALYTIC TOOL FOR QUALITATIVE RESEARCH, QUALITATIVE RESEARCH 1:3, 385-405 (Dec. 2001). I am grateful to Meg Jones for these suggestions.

36 F.2d 1395 (9th Cir. 1992), *cert. denied*, 508 U.S. 951 (1993).

37 *Id.* at 1399.

38 *Id.* at 1396.

had not "knowingly use[d] another's name, voice, signature, photography, or likeness."[39] Relief under the statute is narrow: the term "likeness" is limited to a visual depiction of the plaintiff.[40] Because Samsung's ad "used a robot with mechanical features, and not, for example, a manikin molded to White's precise features," it did not fall within the meaning of California Civil Code section 3344.[41] "However," the Ninth Circuit found, "the common law right of publicity is not so confined."[42]

In finding for White under the common law right to publicity, which also involves the appropriation of a plaintiff's name, picture, or other likeness, the court offered an elaborate hypothetical:

> Consider a hypothetical advertisement which depicts a mechanical robot with male features, an African-American complexion, and a bald head. ... The ad depicts the robot dunking a basketball one-handed, stiff-armed, legs extended like open scissors, and tongue hanging out. Now envision that this ad is running on television during professional basketball games. Considered individually, the robot's physical appearance, its dress, and its stance tell us little. Taken together, they lead to the only conclusion any sports viewer who has registered a discernable pulse in the past five years would reach: the ad is about Michael Jordan.[43]

Similarly, Samsung's depiction of a robot in a gown, jewelry, and blonde wig turning over letters on a game show set could only signify Vanna White. To limit liability on these facts would be, for the majority, to "permit the evisceration of the common law right of publicity."[44]

Judge Alarcon, dissenting, would not have drawn a distinction between the common law and statutory notions of "likeness."[45] For Judge Alcaron, nearly every issue boiled down to the impossibility of anyone confusing the robot for the host. It was "clear that a metal robot and not the plaintiff, Vanna White, is depicted in the commercial advertisement" and indeed "no reasonable juror could confuse a metal robot with Vanna White."[46] The "crude features of the robot are very dissimilar to Vanna White's at-

---

39 *Id.* at 1397.
40 *Id.*
41 *Id.*
42 *Id.*
43 *Id.* at 1399.
44 *Id.*
45 *Id.* at 1402 (Alarcon, J., concurring in part, dissenting in part).
46 *Id.* at 1404, 1405.

tractive human face."[47] Put simply: "One is Vanna White. The other is a robot. No one could reasonably confuse the two."[48]

Samsung's robot generates a tension, evident in the sheer distance between the majority and dissent's respective starting points. For the majority, a robot in these circumstances could not but invoke an individual to anyone with "a discernable pulse."[49] For the dissent, the distinction between a robot and a person is "patently clear."[50] You see hints of the tension in the majority's conflicting interpretation of "likeness" for purposes of statutory and common law, as well as its analysis of likelihood of confusion under the Lanham Act. "One the one hand," noted the majority, "all of the aspects of the robot ad identify White; on the other, the figure is quite clearly a robot."[51]

The Ninth Circuit would confront the issue again just a few years later in *Wendt v. Host International, Inc.*[52] This action was brought by the two actors who played famed barflies Cliff and Norm on the television show *Cheers*.[53] Plaintiffs alleged that the defendant company built two "animatronic robotic figures" or "robots" of them for use in airport bars modeled on the set of *Cheers*, violating their individual rights of publicity by appropriating their likenesses for purposes of the same California state statute at issue in *White*.[54] The robots differed from the clearly metal robot in *White* in that they were embodied, human-looking, and had specific names (Hank and Bob) that differed from those of the plaintiffs.[55] The district court, observing both the robots and the plaintiffs "live" and in person, found them "totally different" and ordered summary judgment for the defendant.[56]

---

47 *Id.* at 1406 (discussing plaintiff's claims under the Lanham Act).

48 *Id.* Judge Alarcon—like Judge Kozinski in *Wendt*—drew a distinction between Vanna White's role as host and her identify as an individual. *Id.* at 1404. Moreover, he thought it was clear from the fact that Samsung was using a robot, rather than Vanna White herself, that she did not endorse their product. *Id.* at 1407.

49 *Id.* at 1399.

50 *Id.* at 1404.

51 *Id.* at 1400.

52 F.3d 806 (9th Cir. 1997).

53 *Id.* at 809.

54 Unlike in *White*, the defendant in *Wendt* received explicit permission from the television show to build the surrounding environment. Thus, plaintiffs proceeded entirely in their personal capacity. *See id.*

55 *Id.*

56 *Id.*

Again the Ninth Circuit reversed. It began by noting that *White* left open the prospect that a "manikin molded to [plaintiff's] precise features" could qualify as a likeness even under the California statute.[57] The appellate court then concluded from *its own* inspection of the robots that a reasonable jury could find them similar enough to violate California law.[58] The court also found likelihood of confusion, in part because people would come up to the plaintiffs and say things like, "Hey George, I just had a drink with you in Kansas City."[59] There was no dissent in *Wendt* but, as alluded to above, Judge Kazinski vehemently dissented from the Ninth Circuit's decision to deny rehearing en banc.[60]

White or *Wendt* are well-known cases of robot impersonation, appearing in textbooks on torts and intellectual property. Less remembered is the earlier case of *Elnicky Enterprises v. Spotlight, Inc.*[61] Rather than a robot copy of a person, *Elnicky* involved a robot copy of another robot. The key question in *Elnicky* was: did the defendant's action of replicating the plaintiff's robot, which he used at trade shows to entertain corporate guests, constitute a violation of the Lanham Act prohibitions on unfair competition?[62]

Throughout its analysis, the *Enlicky* court referred to Rodney and Walter Ego as "he."[63] As in *Wendt*, the judge observed Rodney, the original robot, as well as the usurper Walter Ego.[64] According to the court, "Rodney was casually attired and presented a rakish appearance," whereas "Walter Ego is clean shaven, and has a grin reminiscent of Mortimer Snerd."[65] Rodney was of higher quality and operated with greater skill.[66] And the court went out of its way to note that Rodney's design likely

57 *Id.* at 810.

58 *Id.* ("[W]e conclude from our own inspection of the robots that material facts exist that might cause a reasonable jury to find them sufficiently 'like' the appellants.").

59 *Id.* at 813.

60 Wendt v. Host Int'l, Inc., 197 F.3d 1284 (9th Cir. 1999) (Kozinski, J., dissenting from denial of petition for rehearing). His basis was that actors should not retain intellectual property rights in the characters they play. *Id.*

61 Elnicky Enterprises, Inc. v. Spotlight Presents, Inc., 1981 WL 48202 (S.D.N.Y. 1981).

62 *Id.* at *1.

63 *See, e.g.*, *id.* at *1-2.

64 *Id.*

65 *Id.* at *1, *4. Mortimer Snerd was a puppet used by the legendary ventriloquist Edgar Bergen.

66 *Id.* at *4.

couldn't be patented.[67] Nevertheless, the court found the two robots to be so similar as to confuse potential consumers.

The case made headlines in 1981. The reason had less to do with the difficult competition law question and more with the relief U.S. District Judge Charles Brieant went on to order. The judge wanted Walter Ego dismantled above the torso.[68] This was a strange Lanham Act remedy to say the least, and prompted news stories with headlines like "Robot beheaded" and "Walter Ego loses his head" across North America.[69] The case has since faded with history.

The entire line of robot appropriation cases is interesting for the light it sheds on judicial and public assumptions around robots.[70] In previous work, I have noted that robots contain a unique social valence among artifacts.[71] In psychological studies, for instance, respondents struggle to characterize robots as animate or inanimate.[72] In *White* and *Wendt*, courts are struggling instead with whether a robot version of a person can be said to *represent* that person in the way the law cares about. And in *Enlicky*, the public is reacting to a remedy that feels odd or even wrong given the anthropomorphic qualities of the disputed object, but which would fail even to raise an eyebrow were the artifact a toaster. The purpose of this section was to introduce the general ambiguity that surrounds robots as surrogates; the next few cases explore the issue in greater detail.

---

67 *Id.* at *10 ("In this Court's view, any patent issued for Rodney or any parts of his articulation would be void for obviousness.").

68 *Id.* at *12 (finding that "the dismantling of Walter Ego's head and torso will be required").

69 *See* Robot Beheaded, WILMINGTON MORNING START, Aug. 4, 1981. ("'Off with his head!' a judge has ordered in the case of a robot born of a stolen design."); *Walter Ego loses his head*, THE MONTREAL GAZETTE, Jul. 21, 1981. For a contemporary example of the media's fascination with beheaded robots, see Jack Nicas, *Mannequins Step In For Human Billboards, But Some Are Losing Their Heads*, WALL ST. J., Nov. 30, 2015.

70 Cases in this line continue. In the recent *Brill v. Walt Disney Co.*, for example, a stock car driver sued a motion picture studio for allegedly appropriating his likeness with a driverless car. 246 P.3d 1099, 1103 (2010).

71 *See* Calo, *supra* note 8, at 545-49.

72 Id, at 532 (citing Peter H. Kahn, Jr., et al., *The New Ontological Category Hypothesis in Human-Robot Interaction*, 2011 PROC. 6TH INT'L CON. ON HUMAN-ROBOT INTERACTION 159 (collecting studies)).

II. Robotic Performance

A robot can appropriate the likeness of a performer under certain conditions. But can a robot itself "perform"? The question came up in the context of a tax on entertainment. In *Comptroller of the Treasury v. Family Entertainment Centers*, a Maryland special appeals court had to decide whether life-sized, animatronic puppets that dance and sing at Chuckie Cheese restaurants trigger a state tax on food "where there is furnished a performance."[73] In its analysis, the court looked to Webster's dictionary, which defines performance as a "formal exhibition of skill or talent as a play, musical program, etc.; a show."[74] For the court, it followed that a performance "has connotations of inherent human input that leaves room for spontaneous imperfections during the exhibition of skill or talent."[75]

The court found that, while they "are designed to give the impression that they are performing," the Chuckie Cheese robots fell outside the scope of the statute.[76] In the court's words:

> [A] pre-programmed robot can perform a menial task but, because a pre-programmed robot has no 'skill' and therefore leaves no room for spontaneous human flaw in an exhibition, it cannot 'perform' a piece of music ... Just as a wind-up toy does not perform for purposes of [the statute,] neither does a pre-programmed mechanical robot.[77]

The original tax court also found it noteworthy that the "cyberamic figures" the restaurant chain purchased had yet to be invented when Maryland passed its performance tax statute.[78] Had they existed, the lower court reasoned, surely the legislature would have added them to the list of exceptions, which include "mechanical music, radio, or television, alone."[79] Both the tax and special appeals courts invoked a juke-box as the closest analogy to the robots in Chuckie Cheese.

---

73 519 A.2d 1337, 1338 (Md. 1987). The case was eventually overruled on other grounds involving the question of whether Chuckie Cheese tokens counted as "tickets" for purposes of municipal tax law. *See* 318 North Market Street, Inc. at al. v. Comptroller of the Treasury, 554 A.2d 453 (Md. 1989).

74 *Id.* at 1339.

75 *Id.*

76 *Id.*

77 *Id.*

78 Family Entertainment Centers of Essex, Inc. v. Comptroller of the Treasury, 1985 WL 6106, *1 (Md. Tax 1985).

79 *Id.*

Is a robot an "embellished juke-box"? And would a legislature today include robots in a similar ordinance? Consider the example of *The Robotic Church*, a 2014 art installation by noted sculptor Chico MacMurtrie.[80] MacMurtrie filled a deconsecrated church in the Red Hook neighborhood of Brooklyn with forty or so kinetic sculptures.[81] The sculptures used a wide variety of mechanisms to generate noises and movement at random intervals. The effect is inhuman and surreal, and varies depending on where the audience stands, what robots happen to be working that day, and myriad other factors. It seems unlikely that any visitor could walk away doubting he or she had seen a performance, let alone that the installation was best analogized to a jukebox.

Following the court's reasoning in *Family Entertainment Centers*, MacMurtrie's installation is not a performance because the sculptures have no "skill" and can't exhibit spontaneous flaws.[82] Nor can we say that it is the artist MacMurtrie who is really performing; presumably someone also built the robots in Chuckie Cheese and programmed them to move about in a specific way as well. Consider, too, the ambiguity around the term "preprogrammed." At the time of *Family Entertainment Centers*, robots largely carried out repetitive tasks. By the late eighties, many factories had mechanized every task they could, resulting in an explosion in the number of so-called industrial robots.[83] These machines could only do what they were programmed to do.

Even at the time of *Family Entertainment Centers*, however, roboticists at the Stanford Research Institute (SRI) had developed the famous robot Shakey, capable of basic autonomous actions such as mapping a room and planning a path around an obstacle.[84] Robotics has since moved toward ever greater adaptability. Today's robots "learn" tasks just by watching human demonstrations or even through the robot's own trial and error.[85] The

80 [cite].
81 *Id.*
82 Presumably the word "spontaneous" in this context excludes sudden glitches, for which real robots are famous. A Chuckie Cheese robot that, say, spontaneously combusted would not suddenly be engaged in a performance.
83 [cite].
84 For a nice account of Sharkey, see JOHN MARKOFF, MACHINES OF LOVING GRACE: THE QUEST FOR COMMON GROUND BETWEEN HUMANS AND ROBOTS (2015) 1-7. Robotic art also dates back to the 1960s. *See* Eduardo Kac, *Foundation and Development of Robotic Art*, ART J. 56:3, 60-67 (Autumn, 1997).
85 Calo, *supra* note 8, at 538-39.

robotic warehouses of online retail giant Amazon "organize themselves," such that no human necessarily knows where an item is on the shelf.[86] Artists, including music artists, are aware of robots' new aptitude for spontaneity. They build robots that create art and play alongside robots in live concerts.[87] If ever there were a line between human and robot spontaneity or skill, it is rapidly disappearing.

III. Animate Objects

Closely related, perhaps, is the question of whether a robot represents something "animate" for purposes of tariff schedules. This question has come up repeatedly in American case law, as far back as the 1950s. For historic reasons, taxes on dolls have differed from those on other toys.[88] Tariff law understands dolls—which vary widely in terms of size, materials, and detailing—as distinct from other toys in that dolls represent "animate" life. Thus, the Tariff Act of 1930 drew a distinction between "Figures or images of animate objects, wholly or in chief value of metal" and toys "having a spring mechanism (except figures or images of animate objects)."[89]

In *Louis Marx & Co. and Gehrig Hoban & Co., Inc. v. United States*, a customs court had to decide whether a "mechanical walking robot" being imported represented an animate object.[90] The importers argued that it did, and therefore that the robot should be taxed at 35 "per centum ad valorem" instead of the 50 percent assessed by customs officials.[91] According to the trial transcription, the judge asked counsel for the plaintiff-importers whether the toy was "an imitation of an animate object."[92] Counsel replied: "Yes, a robot. It is as a synthetic man. It is something imitating

---

86 The tagline of Kiva Systems, prior to its purchase by Amazon in 2012 for $775 million, was: "Where products organize themselves." Ryan Calo, *A Robot Really Committed a Crime: Now What?*, FORBES, Dec. 23, 2014, http://www.forbes.com/sites/ryancalo/2014/12/23/a-robot-really-committed-a-crime-now-what/#53862b9a1411.

87 [cite].

88 [cite].

89 19 U.S.C. § 1513 (1930).

90 40 Cust. Ct. 610, 610 (1958).

91 *Id.*

92 *Id.*

men. That is the animate object that this particular toy represents. Also, the common meaning of robot supports our contention."[93]

The court disagreed. As in *Family Entertainment Centers*, the court turned immediately to the dictionary—in this case, two dictionaries.[94] In 1958, Webster's defined a "robot" as "Any automatic apparatus or device that performs functions ordinarily ascribed to human beings, or operates with what appears to be almost human intelligence."[95] Funk & Wagnalls defined a robot as "An automaton that performs all hard work; hence, one who works mechanically and heartlessly."[96] To these, the court compared each dictionary's definition of the word "animate."[97] These were, respectively: "Endowed with life; alive" and "Possessing animal life, living."[98] For the court, there was "nothing in either of the quoted definitions from which to draw the conclusion that a robot is an animate object."[99] A robot is "not a living thing; it is not endowed with life. A robot is a mechanical device or apparatus, a mere automaton, that operates through scientific or mechanical media."[100] The court overruled plaintiffs' exception to the higher tariff.

As much as anything, the court's certainty about its decision in *Louis Marx & Co.* seems noteworthy. There was never any handwringing. And yet, at least one of the dictionaries the court consulted described robots as performing functions "ordinarily ascribed to human beings" and possessing "what appears to be almost human intelligence."[101] The other dictionary referred ambiguously to a robot as "one who works" in a particular way, almost as if to suggest that a robot is a specific type of person.[102]

The court's reasoning was curious in another way: it appeared to assume that the toy being imported—a mechanical walking robot—was not *itself* a robot, but only represented one. According to the court: "A robot is

---

93 *Id.* at 611. Another judge asked about the toy: "It is just a tiny robot?" To which counsel replied, "Yes." *Id.*

94 *Id.*

95 *Id.* (citing *Webster's New International Dictionary*). The 2016 *Merriam-Webster Dictionary* defines robot differently. *See* infra.

96 *Id.*

97 *Id.*

98 *Id.*

99 *Id.*

100 *Id.*

101 *Id.* (citing *Webster's New International Dictionary*).

102 *Id.* (citing *Funk & Wagnalls New Standard Dictionary*).

a mechanical device or apparatus, a mere automaton, that operates through scientific or mechanical media. It, therefore, follows that the toy under consideration, *which simulates a robot*, is not within the statutory language."[103] In other words, although a robot is a machine that simulates a person, a toy robot is only a simulation of the simulacrum. We are left to wonder how robotic a toy must be to itself qualify as a robot.

The same issue arose again in 1971 in *Lewis Galoob Co. v. United States*.[104]Custom officials had assessed a 35 percent duty on a battery-operated Japanese toy called the Swivel-O-Matic Astronaut rather than the 21 percent duty then applicable to toy figures of animate objects.[105] According to the customs court, the Japanese toy consisted of a "representation" of a "mechanical robot" that, when activated, "slides forward on rubber wheels at the bottom of the feet."[106] The toy's chest lit up and opened to reveal two guns that pop out and appear to fire. A chief difference between this toy and that at issue in *Louis Marx & Co.* is the presence of a human face (like an astronaut).[107] Nevertheless, the court had no trouble characterizing the robot as representing something inanimate: "the presence of a human face in an article which is otherwise incapable of representing any living being cannot make 'animate' what is, in totality, incapable of animation or life."[108] Again, the plaintiff's exception was overruled.

By the 1990s, tariff law had changed to mention robots.[109] Interestingly, given the case law, the Harmonized Tariff Schedule characterized robots as "non-human creatures" and lumped them in with toys representing animals.[110] The operative distinction became whether a given item for import constituted a "doll," which in turn required that the toy specifically represent a human being.[111] This, too, proved contentious, as the 2003 case of *Toy Biz, Inc. v. United States* illustrates.[112]

---

103 *Id.* (emphasis added).
104 Cust. Ct. 484 (1971).
105 *Id.* at 485.
106 *Id.*
107 *Id.*
108 *Id.* at 486.
109 *See* Harmonized Tariff Schedule of the United States, 9502.10.00 et seq. (1994).
110 *Id.* at 9503.49.00, HTSUS (1994).
111 *Id.* at 9503.70.90, HTSUS (1994).
112 248 F. Supp. 2d 1234 (U.S. Ct. Int'l Trade 2003.).

In *Toy Biz, Inc.*, the U.S. Court of International Trade had to determine whether figurines of Marvel Comics superheroes and villains were dolls and hence subject to a tariff of 12 percent, rather than the 6.8 percent applicable otherwise.[113] This in turn required the court to decide if the fictional characters, some of which had robotic features, were themselves human. Sometimes the choice was easy: the character Robot Wolverine is obviously a robot and hence a non-human creature.[114] In other instances the call was harder. Is Spider Man not a human being? What about Kingpin, who is just a very bad person?[115]

To decide, the court made the familiar move of consulting a dictionary. The Oxford English Dictionary the court consulted defined a robot as "a machine (sometimes resembling a human being in appearance) designed to function in place of a living agent."[116] More relevant to the court, however, was the OED's definition of a "mutant," which is how Marvel Comics characterized most of the figurines.[117] The OED emphasized that a mutant *begins* as a human but winds up as something else by virtue of a mutant gene.[118] Thus, by definition, mutants are no longer human beings and hence representations of them are not dolls, at least according to the court.

Toy Biz, Inc. helps illustrate how the evolution of human imagination—reflected, for instance, in the evolving definition of the term robot—winds up posing interesting challenges around legal categorization. Again we see evidence of confusion and equivocation. For instance: in a world of prosthetic hearts, ears, arms, and legs, what do we make of the court's contention that "robotic features, such as artificial eyes or limbs," militate against a finding of humanity?[119] And why would one such robotic feature

113 *Id.* At 1240.

114 *Id.* at 1241 n.10 (discussing imports that the parties stipulated to be non-human).

115 *Id.* at 1252 ("Nothing in the storyline indicates that Kingpin possesses superhuman powers.").

116 *Id.* at 1251 (citing 9 OED 1036-37 (2d ed. 1989)). Today the OED has a long definition of robot that encompasses everything from a "machine capable of automatically carrying out a complex series of movements, esp. one which is programmable" to a "central European system of serfdom, by which a tenant's rent was paid in forced labour or service" to a "traffic light." [cite].

117 248 F. Supp. 2d, at 1251 (citing 10 OED at 145-46).

118 *Id.*

119 *Id.* At 1251.

convert a person into a robot when a human face does not convert a robot into an astronaut?[120] I discuss these issues in greater detail in Part III.

IV. Robot Possession

The case studies I've presented thus far interrogate the degree to which robots resemble people. The next two case studies involve a different question, namely, whether robots can be thought of as extending people into physical space in ways the law cares about. The contexts—maritime law of salvage in this section and state criminal law of burglary in the next —are quite disparate. But there are nevertheless some interesting similarities.

Salvage rights to a famous shipwreck were at issue in the next robot case, *Columbus-America Discovery Group, Inc. v. The Unidentified, Wrecked, and Abandoned Vessel, S.S. Central America.*[121] The *S.S. Central America* was a steamship that sank in the Atlantic Ocean in 1857 carrying gold from the California Gold Rush.[122] Many looked for the *Central America* following its accident; no one could find it. One hundred and thirty years later, a high tech operation discovered the wreckage and claimed first salvage rights at maritime law.[123] Columbus-America Discovery Group made use of some very new technology for the late 1980s, including an unmanned (i.e., robotic) submersible equipped with cameras and actuators capable of grasping objects at the direction of its operators.[124]

The trial court in *Columbus-America* had to decide whether, for purposes of salvage law, Columbus-American Discovery Group "achieved exclusive custody, control, and possession of the wreck."[125] First salver rights at maritime law entitle the operation to recover some substantial portion of

120 *See supra*, note 105 and accompanying text.

121 1989 A.M.C. 1955 (1989).

122 Columbus-America Discovery Group, Inc. v. The Unidentified, Wrecked and Abandoned Vessel, S.S. Central America, 742 F.Supp. 1327, 1328-29 (E.D. Va. 1990) (described the facts around the shipwreck and salvage operation), rev' on the other grounds by Columbus-America Discovery Group, Inc. v. Atlantic Mut. Ins., 974 F.2d 450 (4th Cir. 1997).

123 *Id.*

124 *Id.* at 1329.

125 *Id.* at 1330.

the treasure aboard the vessel as well as exclude other potential salvers.[126] And indeed, several other teams were in the area searching for the *Central America*, leading the Columbus-America Discovery Group to ask the court to enjoin anyone else from entering the immediate salvage zone.[127]

The usual way for custody, possession, and control to be achieved at this time was by human divers approaching the vessel and either recovering property over time or, if safe, lifting the wreck out of the water.[128] The salvage team in *Columbus-America*, however, was not able (or willing) to send anyone that far down—nearly one and one half miles below the surface. It sent down its robots instead.[129]

The court decided that, in light of the conditions, sending the robots counted for purposes of effective control and possession. They were, after all, able to generate live images of the wreck and had the further "capability to manipulate the environment" at the direction of people.[130] The court fashioned a new test for effective possession through "telepossession," consisting of four elements: (1) locating the wreckage, (2) real-time imaging, (3) placement of a robot near the wreckage with the ability to manipulate objects therein, and (4) intent to exercise control.[131] As maritime law scholar Barlow Burke, puts it: "This is as close as the court can come to creating a new legal basis for establishing possession without actually doing so."[132] On the basis of the new test, which has been cited by other courts since, the court granted salvage rights to Columbus-America Group and enjoined its competitors.[133]

---

126 Columbus-America Discovery Group, Inc., 1989 A.M.C. at 1957-58.

127 *Id.*

128 *See* Barlow Burke, Jr., *A Reprise of the Case of Eads v. Brazelton*, 44 ARK. L. REV. 425, 456-58 (1991).

129 Columbus-America Discovery Group, Inc., 742 F. Supp. at 1331 (noting that "salvage operations will ultimately be effected by the use of a remotely operated vehicle capable of handling the remains of the vessel and retrieving its contents").

130 Columbus-America Discovery Group, Inc., 1989 A.M.C. at 1958.

131 *Id.* at 1957-58.

132 *See* Burke, *supra* note 128, at 456.

133 R.M.S. Titanic, Inc. v. Wrecked & Abandoned Vessel, involving what is perhaps the most famous shipwreck in modern history, cites the Columbus-America test. 742 F. Supp. 2d 784, 794 (E.D. Virginia 2010). For another example of a case adopting the doctrine of telepossession, see Ocean Mar, Inc. v. The Cargo of SS Islander, 1998 WL 886109 (D. Alaska Aug. 28, 1998).

The 1990s saw a renaissance of deep sea treasure hunting, in large part due to advances in robotics.[134] Today the use of sea and undersea robotics is even more extensive. Both the public and private sector are making increasing use not only of teleoperated robots, i.e., machines under the constant direction of a remote pilot, but autonomous systems that explore the sea on their own.[135] Navy pilotless submarines monitor undersea activity.[136] And private or academic research vessels spend months on the open waves mapping out territory. The company Liquid Robotics lists over 625,000 autonomous miles at sea with its Wave Glider robots.[137]

It is interesting to note that *Columbus-America* and its progeny leave open the question of whether the *autonomous* discovery of a shipwreck could ever support a first salvage claim. There is reason to believe it might not. First, autonomous exploration does not technically meet the *Columbus America* elements, which emphasize the role of a human operator above water and require intent to exercise control.[138] A court might not credit an autonomous submarine with the "intent" to record or manipulate a wreck, even were it physically able to do so. Second, the court's new standard grew from the recognition that high-seas salvage operations were dangerous, even if the crew did not go underwater.[139] This leaves open the prospect that an all-robot expedition would not meet the spirit of so-called telepossession because the human crew would be in safer waters or even on shore.[140]

---

134 [cite].

135 R.M.S. Titanic, Inc., for instance, involved a mixture of teleoperated and autonomous submarines. 742 F. Supp. 2d at 799, n.20.

136 *See* Craig H. Allen, *The Seabots are Coming Here: Should they be Treated as 'Vessels'?*, 65 J. NAVIGATION 749 (2012).

137 *See* http://www.liquidr.com/ (last visited Feb. 17, 2016).

138 Specifically, the fourth factor requires "present intent to control ... the location of the object." Columbus-America Discovery Group, Inc., 1989 A.M.C. at 1958.

139 *Id.* (finding that "Columbus-America has maintained a reasonable presence at the cite," given the "special circumstances which characterize deep ocean salvage, including rough seas, sailing distances to safe port, remoteness from repair facilities and assistance...").

140 Obviously the stakes are entirely different, but this question mirrors in a sense contemporary arguments around the propriety of using drones or other robots to kill at distance without imperiling American soldiers. For a discussion, see Singer, *supra* note 18, at 309-14 (discussing perceptions of the United States' use of remote warfare).

V. The Robot Burglar

In 1887, an English court convicted Henry Hands of common law larceny for his appropriation of several cigarettes belonging to an Edward Shenton.[141] Despite somewhat unusual facts for the time, neither the lower court nor the court on appeal had any trouble characterizing Hands' actions as theft. What Hand and his accomplices did was use a brass and lead disc instead of a penny to retrieve cigarettes from Shenton's new "automatic box."[142] According to the court, Hands' substitution of a disc without value for a penny defrauded Shenton's machine—"the means ... were fraudulent"—and thereby deprived Shelton of his cigarettes without his consent.[143]

Reg. v. Hand and Others involved theft from a machine. The case I want to examine here is the slightly more fanciful prospect of theft *with* a machine. Robots make another appearance in *People v. Davis*, a 1998 burglary case before the California Supreme Court.[144] The state accused Michael Wayne Davis of passing a bad check to a window teller through a chute. The teller grew suspicious and phoned the police, who picked up Davis while he was still waiting for the bank to cash the fraudulent check.

As in *Hand*, the trial and appellate courts in *Davis* had no problem finding larceny on these facts.[145] But one of the charges was burglary, i.e., entering a building or structure with the intent to steal. Relying on an earlier appellate decision in *People v. Ravenscroft*, the lower court found that Davis' actions were, in fact, a burglary as well.[146] *Ravenscroft* involved the burglary of an ATM through the use of a stolen card.[147] Other California courts had similarly held that any wrongful entry into a structure, how-

---

141 Reg. v. Hand and Others, LVI. Cr. Cas. Res. 370 (May 14, 1857). At English common law, larceny had similar elements to statutory theft today; the defendant must intentionally deprive another of goods of value without permission. *See* Minturn T. Wright III, Note, *Statutory Burglary—The Magic of Four Walls and Roof*, 100 U. PENN. L. REV. 411 (1951). Thank you to James Grimmelmann for bringing *Reg. v. Hand and Others* to my attention.

142 Reg., Cr. Cas. Res. at 370.

143 *Id.* at 371.

144 958 P.2d 1083 (Cal. 1998).

145 *Id.* at 1084.

146 *Id.* at 1087.

147 *See* People v. Ravenscroft, 198 Cal. App. 3d 639 (1988).

ever slight or partial, could support a burglary charge.[148] The dissent in *Davis*, like the lower court, would have found burglary in the act of approaching the security window and passing through it "an instrumentality to trick the teller into handing him money back."[149]

The majority was of another view; it abrogated *Ravenscroft* and rejected the burglary charge.[150] The court first conceded that a defendant could commit burglary without himself entering the premises—for instance, through the use of a robot. "Instruments other than traditional burglary tools certainly can be used to commit the offense of burglary."[151] For example, "a robot could be used to enter the building."[152] But "it does not necessarily follow from these conclusions that insertion of a stolen card into an ATM constitutes a burglary."[153] A robot represents a new means by which to accomplish the traditional ends of entry for the purposes of theft. The introduction of a card or check into a structure does not become entry merely because the defendant has a bad intent.

There is a certain intuitive appeal to the majority's reasoning. Passing a bad check through a chute does not feel analogous to entering a facility with a robot without the permission of the owner. Yet the line between a check and a robot may not be as sharp as the majority in *Davis* assumes. The line is not, for instance, that a robot enters the building and a check doesn't. Both enter the building, as the dissent pointed out.[154] And the owner of the building probably would no more consent to bad checks entering his or her bank than bad robots.

Nor is the line that a robot is always an invasion. The court envisions a robot entering a building that the robot's operator is forbidden to enter—a

148 958 P.2d at 1086-87 (citing cases).
149 *Id.* at 1091(Baxter, J., dissenting).
150 *Id.* at 1090.
151 *Id.* at 1087.
152 *Id.*
153 *Id.*
154 *Id.* at 1092, 1099. The dissent proposed that a burglary occurs whenever there is a burglar present at the scene. *Id.* at 1093 ("Simply put: no burglar at the crime scene, no burglary."). The majority invoked robots once again in criticizing this standard: "Under the dissent's proposed rule, a person who used a remote controlled robot, operated from across the street or across town, to enter the building for purposes of committing larceny or any felony would not commit burglary." *Id.* at 1090, n.7.

bank after hours—and extracting property.[155] But imagine instead that an individual visited a bank during business hours with a telepresence robot like those commercially available today.[156] Presumably no one would accuse someone with an immune deficiency of burglary by seeing about a loan by robot instead of in person.

Nor is the line that a human intermediary has to assist with the check by cashing it but not with the robot. Again, if a person were to "enter" a bank during business hours by telepresence and trick or coerce the clerk into handing over money—for instance, by pretending to be another customer or hinting at the presence of a bomb—presumably the *Davis* court would uphold a burglary charge despite the cooperation of the staff.

An even more interesting question arises if the robot is already inside the premises. Personal and service robots represent a significant and growing segment of the exploding robotics market.[157] These new robots are smarter and more versatile in part because they are connected to the Internet—a model James Kuffner calls "cloud robotics" in reference to the idea that the robot's intelligence is somewhere other than its body.[158]

A few years ago, security researchers at the University of Washington showed how easy it is to take over an Internet-enabled robot remotely.[159] The researchers could not only record audio and video, but could also move the robot around the household.[160] Imagine a thief were to take control of a robot already in the home and use it to drop an expensive item—car keys or jewelry—out of the mailbox. This activity would violate fed-

---

155 *Id.* at 1087, 1090, n.7.

156 A telepresence robot allows a person to control and robot remotely and navigate a space via computer. Often there is a video of the operator displayed on the "face" of the robot. The operator can also see or hear the environment and move around in space. *See, e.g.*, https://suitabletech.com/beam/.

157 The International Federation of Robotics keeps statistics on world trends with respect to robotics, which it breaks down according to category. Historically industrial robotics has comprised the bulk of the sector. In recent years, however, personal and service robots have gained a greater share. [cite] *See also*, Calo, *supra* note 8, at 526-28 (citing additional statistics).

158 *See* Erico Guizzo, *Cloud Robotics: Connected to the Cloud, Robots Get Smarter*, IEEE SPECTRUM (Jan. 24, 2011).

159 Tamara Denning, et al., *A spotlight on security and privacy risks with future household robots: Attacks and lessons, in: Proceedings of the 11th International Conference on Ubiquitous*, COMPUTING, Sept. 30–Oct. 3, 2009.

160 *Id.*

eral laws against hacking.[161] But is it a burglary? The robot had permission to enter the facility; the owner placed it there. But the thief did not have permission to enter the robot.[162]

VI. The Unreliable Robot

This Part's final case study involves what we usually think of as the core interaction between robotics and law: legal liability for robot mishaps. This work can tend toward the fanciful and will often take the form of speculation. As David Vladeck, Patrick Hubbard, and others argue, however, some combination of tort law and safety regulations seems well-positioned to address physical harm by robots, at least in the near-term.[163]

I mostly agree with the Vladeck and Hubbard view with two exceptions. First, I believe the new collaborative ecosystem in robotics will shortly confront courts and lawmakers with a novel issue.[164] Someone will have to decide whether manufacturers of robots that are increasingly designed to run third-party code (colloquially, "apps") will be liable when that code leads to physical harm.[165] Federal law immunizes Internet and mobile platforms for what users say and do on those platforms on the theory that it is the user, not the platform, who "publishes" the relevant content.[166] But courts and Congress might decide to strike a different balance

---

161 *E.g.*, Computer Fraud and Abuse Act of 1986, 18 U.S.C. § 1030 (1986) (prohibiting, inter alia, unauthorized access to a protected computer).

162 Holographic projects raise an interesting question as well. In July of 2015, a music performer appeared by hologram at a concern in Hammond, Indiana from a studio in Los Angeles. As he was wanted on an arrest warrant in Indiana, the local police shut down the venue, telling concert goers that his virtual presence posed a safety risk. *See* Daniel Rivero, *This rapper's hologram is changing the way we think about protest*, FUSION, Sep. 15, 2015.

163 *See* David C. Vladeck, *Machines Without Principles: Liability Rules and Artificial Intelligence*, 89 WASH. L. REV. 118 (2014); F. Patrick Hubbard, *'Sophisticated Robots': Balancing Liability, Regulation, and Innovation*, 66 FLA. L. REV. 1803 (2014).

164 Calo, *supra*, note 8, at 532-37 (discussing embodiment).

165 *Id.*

166 *See* Communications Decency Act of 1996, 47 U.S.C. § 230(c)(1) (2006) ("No provider of an interactive computer service shall be treated as the publisher or speaker of any information provided by another information content provider.").

when bones and not bits are on the line.[167] Second, I believe robots will present courts with yet another opportunity to rethink proximate causation. The prospect that robots will behave in ways that are not foreseeable to the designer or user is probably closer than many legal scholars admit.[168] Part III explores this issue in greater detail.

For now, the role of the robot in cases involving physical harm is largely incidental. Every year a robot in a U.S. factory kills one or two people, according to statistics kept by OSHA.[169] Some of these deaths result in lawsuits.[170] There are also medical malpractice or product liability lawsuits that involve robotic surgery.[171] While some of the injuries people sustain would not occur with, for instance, comparable laparoscopic surgery, these cases do not seem to require a reexamination of tort doctrine.[172] We might expect more such cases as drones and other robotic products saturate the market. But as the Introduction makes clear, these and other incidental robot cases are outside the scope of this particular project.[173]

Here I want to concentrate on a narrow category of harms that involve humans relying on robotic systems to their or another's detriment. A relatively straightforward example is the 1949 case of *Brouse v. United States*.[174] In *Brouse*, a federal district court had to decide whether the government could be held responsible for a plane crash between an army fighter and small private plane. "It was clearly established," according to the court, "that the army plane at the time of the collision was under robot control.[175] The court's rejection of the government's position in *Brouse*—that the collision could not have been avoided through reasonable diligence—was emphatic. The court characterized the pilot's obligation "to

---

167 Calo, *supra*, note 8, at 532-37. *See also* Ryan Calo, *Open Robotics*, 70 MD. L. REV. 571 (2011).

168 Calo, *supra*, note 8, at 538-45 (discussing emergence).

169 John Markoff and Claire Cain Miller, *As Robotics Advances, Worries of Killer Robotics Rise*, N.Y. TIMES, Jun. 16, 2014.

170 *E.g.*, Payne v. ABB Flexible Automation, Inc., 116 F.3d 480 (8th Cir. 1997).

171 *E.g.*, Reece v. Intuitive Surgical, Inc., 63 F. Supp. 1337 (N.D. Ala. 2014).

172 One possible exception is a case wherein an individual sued a hospital because it chose not to use a robot for surgery and to rely instead on a human-only surgery. Mracek v. Bryn Mawr Hosp., 2010 Lexis 2015 (3rd Cir. Jan. 28, 2010).

173 [cite].

174 83 F. Supp. 373 (N.D. Ohio 1949).

175 *Id.* at 374.

keep a proper and constant lookout" as "unavoidable" and found the plaintiff's right to recovery on the facts to be "without question."[176]

Of course, robotic control today looks nothing like it did at time of the accident; in 1947, autopilot consisted of mechanical tension rods that merely kept the plane flying on the same path.[177] By 1994, navy fighter jets were landing on aircraft carriers without human input.[178] Nevertheless, as David Vladeck argues, "In cases involving other autonomous machines, liability has been difficult to establish where alternative theories of liability are present" that implicate a person's conduct.[179] Vladeck's main example is *Ferguson v. Bombardier Services Corp.*[180] *Bombardier Services Corp.* is a relatively recent case in which the court rejected liability for the manufacturer of an autopilot system. The system was engaged at the time of the crash and might have explained the crash.[181] But the court decided against the airline instead on the theory that the plane had been improperly loaded.[182]

The tendency of courts to locate liability for robots in people is subject to several caveats. First, as the Third Circuit observed in the context of a robotic practice pitcher that threw wild, "robots cannot be sued."[183] Thus, the question is not whether the robot will be liable, but only *which* person will be liable. Will it be the manufacturer of the autopilot system or surgical robot, or the pilot and the surgeon?

Second, we do not necessarily see the same tendency in the absence of physical harm. Take the example of *Royal Insurance Company of America v. Crowne Investments*.[184] In *Royal Insurance*, the Supreme Court of Alabama had to decide whether to uphold a default judgment in favor of Crowne Investments. Crowne had served process to Royal but, according

176 *Id.* at 374-75.

177 [cite].

178 Philip E. Ross, *When Will We Have Unmanned Commercial Airliners?*, IEEE SPECTRUM (Nov. 29, 2011) (interview with roboticist and former navy pilot Missy Cummings), http://spectrum.ieee.org/aerospace/aviation/when-will-we-have-unmannedcommercial-airliners.

179 David C. Vladeck, *Machines Without Principals: Liability Rules and Artificial Intelligence*, 89 WASH. L. REV. 117, 140 (2014).

180 *Id.* at 140, n.78 (citing 244 F. App'x 944 (11th Cir. 2007)).

181 *Id.*

182 *Id.*

183 United States v. Althone Industries, Inc., 746 F.2d 977, 977 (3rd Cir. 1984).

184 903 So.2d 802 (Ala. 2004).

to Royal staff, the relevant email was lost due to a glitch in Royal's "robotic mail system."[185] The court could have followed the reasoning of *Brouse* and subsequent cases that people have a responsibility to monitor automated systems and risk being held accountable if they do not. Instead, the court declined to find that the default judgment. For the court, the glitch was not "a result of the defendant's own culpable conduct,"[186] and Royal was "in no way culpable" for its robotic mail system.[187]

B. ROBOTS AS LEGAL SUBJECTS

The first Part of this article offers a series of case studies as a corrective to the intuition that all interesting robot cases wait in the future. It hopes to illustrate that, while often the role of the robot is incidental to the underlying legal problem of a case, robots throughout the decades occasionally present courts with quite interesting or challenging issues. The way courts puzzle through them may shed some light on how the law may react to the rapid mainstreaming of robots we see today.

This second Part discusses another sort of case—a case in which no robot is at issue, except in the imagination of the judge. Reading through hundreds of cases, one is struck by the frequency and manner in which courts invoke robots to make observations about people or law. Humans are, or are not, like robots, a critical distinction that informs the legal issue before the court. What emerges is a fairly clear and consistent picture of a robot in the judicial mind: robots may appear to be agents or entities but in actuality are only tools. Robots are defined precisely by reference to their complete lack of discretion or capacity for spontaneity or judgment.

It is fair to ask what, if any, lessons we can draw from the way judges talk about a given technology, especially when that technology is not be-

185 *Id.* at 806.

186 *Id.* at 808 (Ala. 2003) (citing Kirtland v. Fort Morgan Auth. Sewer Serv. Inc., 524 So.2d 600, 605 (Ala. 1988)).

187 *Id.* at 812. The court acknowledged that Royal was "at worse negligent." *Id.* In rare instances, courts may also hold parties accountable when a software problem results in injury to property or even economic loss. *E.g.*, Pompeii Estates, Inc. v. Consol. Edison Co. of N.Y., Inc., 397 N.Y.S. 2d 577 (N.Y. Civ. 1977) (service termination following a computer glitch). But pure information-based harms are generally not compensated. *See* Ryan Calo, *Open Robotics*, 70 MD. L. REV. 598-61 2010 (furnishing examples).

fore the court. Several strains of research suggest the importance of mental models to legal outcomes. As an initial matter, rhetorical allusions appear capable of shaping policy recommendations. In 2011, Paul Thibodeau and Lera Boroditzsky at Stanford University conducted an experiment (n = 1,482) in which they presented subjects with a description of an imaginary city experiencing a surge in criminal activity.[188] To one set of subjects, the researchers described crime in general as a "virus infecting the city" and "plaguing" neighborhoods.[189] To the other, they described it as a "wild beast preying on the city" and "lurking in neighborhoods."[190] When asked for policy recommendations, subjects in the first condition recommended more enforcement 56% of the time and social reforms 44%.[191] Subjects in the second condition recommended enforcement and reform 75% and 25%, respectively.[192]

Indeed, judges rely on metaphor and analogy when reasoning through the protection law should afford to new technologies. In the context of cryptography, for instance, Michael Froomkin explores the four metaphors that seem to have the most appeal to the courts: encryption is like a "car" that carries information, a kind of "language," a "safe" that hides secrets, or a "house" in which conversation takes place.[193] According to Froomkin, a judge's selection of metaphor in turn reveals the level of First and Fourth Amendment protection the judge is willing to apply. If encryption is merely a car in which messages travel, it gets lesser constitutional protection.[194] But if encryption is itself a language, it may be protected quite fully. Froomkin's conclusion is that "ideas are weapons."[195]

Judges appear from remarks and arguments to possess a highly homogenous mental model of what a robot is. It turns out not to be a particu-

---

188 Paul H. Thibodeau and Lera Boroditsky, *Metaphors We Think With: The Role of Metaphor in Reasoning*, PLOS ONE 6(2): e16782 (2011).

189 *Id.* at 3-4.

190 *Id.* at 3.

191 *Id.*

192 *Id.* at 6.

193 A. Michael Froomkin, *The Metaphor is the Key: Cryptography, the Clipper Chip, and the Constitution*, U. PA. L. REV. 709, 861-62 (1995).

194 *Id.* at 879.

195 *Id.* at 843 (title of Part IV). *See also* Orin S. Kerr, *The Problem of Perspective in Internet Law*, 91 GEO. L.J. 357 (2003) (arguing that courts come to different conclusions about the scope of a warrant or other Fourth Amendment issues depending on whether they take an "insider" or "outsider" view of the technology).

larly accurate one: The puzzles and potential mistakes that arise in Part I arguably make more sense as we build out the typical judge's mental model of robots. And the way judges talk about robots, catalogued here for the first time in the literature, provides fodder for future investigations of robotics law and policy, which are the subject of Part III. But ultimately I acknowledge that a judge may invoke robots in one way but decide robot related cases in another.

I. The Robot Judge

Robots appear repeatedly in discussions of judicial bias. Confronted with a variety of allegations, many opinions remind us that judges are flesh and blood people. Litigants may *expect* judges to be robotic, but they are not. And nor should they be. We would not want to dispense with human judgment. As one court put it: "We have not, and hopefully never will reach the stage in Alabama at which a stone-cold computer is draped in a black robe, set up behind the bench, and plugged in to begin service as Circuit Judge."[196]

This claim arises in at least two contexts. The first is the reaction of trial judges to the presentation of evidence or other behavior. Perhaps a judge laughs with a funny witness or betrays emotion at the plight of the victim, or displays impatience over delay or interruption.[197] Opposing counsel seizes on this moment in an effort to show the judge is prejudiced against their client. Courts almost universally reject these challenges, often citing to the holding of *Allen v. State* that "the trial judge is a human being, not an automaton or a robot."[198]

Even in the absence of a reaction from the bench, litigants may question whether the very identity of a jurist suggests partiality. A fascinating and historically important case is that of *Commonwealth of Pennsylvania et al.*

---

196 Allen v. State, 290 Ala. 339, 342 (1973).

197 *E.g.*, Keppel v. BaRoss Builders, Inc. 7 Conn. App. 435, 509 (1986) ("Above all, it showed that a judge is a human being, not the type of unfeeling robot some would expect the judge to be."); Fletcher v. State, 291 Ala. 67, 69 (1973) ("[T]he trial judge is not required to be a robot without emotional reaction to happenings in his court. Impatience with excessive delay by counsel is a natural and understandable reaction.").

198 290 Ala. at 342.

v. Local Union 542 et al. from 1974.[199] *Local Union 542* involved allegations of racial discrimination by twelve black workers against a predominantly white labor union.[200] The union sought to disqualify the federal district court who was assigned to the case on the basis that he was himself black and had recently addressed a group of black historians, at which time the judge allegedly displayed "an intimate tie with and emotional attachment to the advancement of black civil rights."[201]

In rejecting the challenge to his impartiality, Judge Higginbotham noted that white judges were free to pursue their own interests and concerns outside the bench; no one expected white judges to renounce their heritage or history to maintain impartiality.[202] The union's recusal motion implied that black judges, in contrast, must be "robots who are totally isolated from their racial heritage and unconcerned about it," or at least refrain from discussing that heritage.[203] "Should they be robots?" Judge Higgonbotham asked of black judges; no more so than any other jurist.[204]

A second context in which we see the specter of the robot judge is in discussions of judicial discretion. A judge need not, for instance, "robotically recite" every statutory consideration in the course of meting out a sentence under the federal sentencing guidelines,[205] or "recite robotic findings" to establish that conditions have changed in a given country for purposes of rejecting asylum.[206] American law follows from a code, but not

---

199 Commonw. of Penn. and Raymond Williams et al. v. Local Union 542, Int. Union of Op. Eng., et al., 388 F. Supp. 155 (1974).

200 *Id.* at 163, n.7.

201 *Id.* at 157.

202 *Id.* at 165.

203 *Id.* at 178.

204 *Id.* at 180. In Local Union 542, the court believed that the petitioners expected black judges to be more robotic than white ones. Meanwhile, in *White v. Samsung*, the Ninth Circuit chose a hypothetical about a famous black athlete to illustrate just how close robots can come to the people they depict. For a further discussion, see *infra* Part III.A.

205 United States v. Ruiz-Salazar, 785 F.3d 1270, 1272-73 (8th Cir. 2015). Other cases dispense the requirement to make "robotic incantations that each factor has been considered." *See, e.g.*, United States v. Blackmon, 662 F.3d 981, 988 (8th Cir. 2011) (citing United States v. Lamoreaux, 422 F.3d 750, 756 (8th Cir. 2005)).

206 Hoxhallari v. Gonzalez, 468 F.3d 179, 187 (2d Cir. 2006).

in the sense of software code that a judge executes like a computer.[207] We assume a judge has considered relevant factors unless we have evidence to the contrary.

Also interesting for our purposes is the observation that appellate courts do not, by reversing and remanding a decision, turn the trial court into *their* robot. Obviously a decision by a higher court "severely limits the kinds of considerations open" to a lower court on remand.[208] At the same time: "An appellate mandate does not turn a district judge into a robot, mechanically carrying out orders that become inappropriate in light of subsequent discoveries or changes in the law."[209] A lower court judge interested in some wiggle room might say that the mere fact she was overturned does not mean she has lost all humanity or judgment.

The intuition may go deeper still, beyond the individual judge. There is a general sense among many litigants and some courts that an overly robotic judicial *system* is not a fair one. In challenging health regulations, for instance, appellants in *Kirk v. Secretary of Health and Human Services* argued that the new guidelines "robotize the adjudicative process, in violation of due process guarantees."[210] Although the court in *Kirk*, ultimately rejected appellants' claim, a similar argument has gotten traction in other contexts such as disability rights.[211] Though we are famously a government of laws, and not of men,[212] those laws are to be interpreted and applied by real men and women.

---

207 Cf. John Greenman, *On Communication*, 106 MICH. L. REV. 1337, 1375 (2008) (advancing a free-will theory of the First Amendment that disputes computer code is speech).

208 Anand v. Nat'l Republic Bank of Chicago, 239 B.R. 511, 520 (N.D. Ill 1999).

209 *Id.* (quoting Barrow v. Falck, 11 F.3d 729, 731 (7th Cir. 1993)). *See also* Jianli Chen v. Holder, 703 F.3d 17 (1st Cir. 2012) ("This multifaceted role is not meant to be robotic. The [Board of Immigration Appeals] is not bound to parrot the precise language used by the [immigration judge] but, rather, may use its own vocabulary.".).

210 Kirk v. Sec'y of Health & Human Serv., 667 F.2d 524, 531 (6th Cir. 1981).

211 E.g., Stewart v. Harris, 508 F. Supp. 345 (D.N.J. 1981). *See also* Jianli Chen v. Holder, 703 F.3d 17 (1st Cir. 2012) (noting that the role of the immigration appeals board "is not meant to be robotic" and that the board has the "prerogative—indeed the duty—of examining the basis for, and then synthesizing and analyzing, the [immigration judge's] findings.".).

212 John Adams, *Letters of Novanglus*, BOSTON GAZETTE (1774).

II. The Robot Juror or Witness

The judge is not a robot and neither is the quintessential finder of fact, the juror. It is not necessarily evidence of bias for a juror to laugh or cry during trial and our Constitution requires courts to scrutinize the bases upon which litigants strike jurors from service.[213] Courts also assume jurors who do serve are people with lived experience, not programmable machines.[214] Thus the court invoked robots in *Burch v. Reading Co.*, a case in which a widow with two children broke her ankle on the job.[215] The judge's instructions were adequate because jurors "are not robots who come to the court house with minds tabula rasa and who respond mechanically to every impression they receive in the courtroom."[216]

People are not robots. When they act like robots in court, this is considered to be a bad sign. Several cases hold robotic behavior in court against the litigant. It may be that cold, calculating people are not viewed as trustworthy; that truth cannot be rehearsed; or that justice is somehow an intrinsically humanistic process.[217] Regardless, we see this sentiment in cases, such as *Rong Lin v. Mukasey*, where a witness appeared to be "robotically repeating a script rather than testifying from actual experience."[218] We also see it in cases, such as *Kung Lin Chen v. U.S. Department of Justice* in which the witness testified well enough but "appeared 'robotic' when pressed for details on cross examination."[219] Such a finding—often,

---

213 *See, e.g.*, Batson v. Kentucky, 476 U.S. 79 (1986) (holding that striking jurors solely based on their race violates the Equal Protection Clause of the Fourteenth Amendment). Courts also grant certain latitude to spectators at a trial in displaying emotion. *See generally* Meghan E. Lind, Comment, *Hearts on Their Sleeves: Symbolic Displays of Emotion by Spectators of Criminal Trials*, 98 J. CRIM. L. & CRIMINOLOGY 1147 (2008).

214 *See* Burch v. Reading, Co. 240 F.2d 574, 577 (3d Cir. 1957).

215 *Id.*

216 *Id.*

217 I would pause to note that the standard could be quite different for experts. No case we came across sought to detract from expert testimony on the basis that it felt coached or robotic. Rather, the issue came up in contexts, like asylum cases, where you might expect emotion intensity. Litigants can also try to leverage previous robotic behavior in court for advantage, as when a defendant attempts to withdraw a guilty plea that he delivered "like a robot." United States v. Osei, 679 F.3d 742 (8th Cir. 2012).

218 299 F. App'x 10 (2d Cir. 2008).

219 195 F App'x 10, 11 (2d Cir. 2008).

by an immigration authority—can support an adverse finding as to credibility.[220] Apparently, testifying in court requires some measure of spontaneity, much like a "performance" for purposes of entertainment taxes on food.[221]

III. The Defendant's Robot

The previous two sections focus on behavior taking place within the judicial system. Courts also invoke robots in considering whether to hold parties accountable for conduct that landed them in the system in the first place. Stated simply, the defendant in this scenario describes him or herself as a robot under the control of some operator not before the court. Alternatively, the actions of an entity such as a corporation can be attributed to the defendant because the entity is simply an extension of the person—his or her robot. The idea of a robot becomes synonymous with the absence of autonomy or free will.[222]

In the 1950 case *Frye v. Baskin*, the plaintiff owned a Jeep that he taught his minor son to drive:[223] His son John was on a date with the defendant, a minor girl, whom John asked to take the wheel. She did not know to drive how but, "under his tutelage," she managed to drive the car around town for a time.[224] At one point, John called out a direction to the girl and then reversed himself, telling her to go right instead of left. She tried to comply and wound up crashing the vehicle.

In the resulting suit by the father against his son's friend, the court refused to find the defendant negligent as a matter of law. According to the

---

220 That the cases involving robotic witnesses tend to arise in the context of immigration could be a function of the fact that an immigration court originally came up with the formulation. Or it may reflect something else, such as a language or cultural barrier. *See* infra notes __ to __ and accompanying text (briefly discussing race and gender dimensions to robotics law).

221 *See supra*, Part I.B.

222 A note about scope: clearly the notion of free will in philosophy and law has a contentious and involved history. I will not attempt here to unpack whether such a concept exists or what its role should be in legal discourse. Rather, I only call attention to the ways litigants and jurists invoke the concept of robot—apparently, an entity human in appearance but lacking free will—in an effort to avoid or pass through culpability for wrongdoing.

223 S.W. 2d 630 (Mo. App. 1950). .

224 *Id.* at 633.

court, plaintiff's son John was really the driver.[225] The defendant "controlled the car the same as if she had been a robot or an automaton. When John said 'turn,' she turned, mechanically."[226] She was merely "the instrumentality by which John drove the car."[227] Accordingly, "if it were negligence, it was John's and not hers."[228] Or at least the jury was entitled to so hold.[229]

Molko and Leal v. Holy Spirit Association involved an allegation of false imprisonment against the Unification Church that came before the Supreme Court of California.[230] Plaintiffs claimed inter alia that they were held captive through brain washing until each were captured and "deprogrammed" by professionals sent by their respective parents.[231] Robots do not come up in the majority opinion as such, which allowed the false imprisonment claim to go forward. But the dissent thought the distinction between people and robots to be crucial: "The evidence before us ... clearly indicates that the Church's indoctrination did not render appellants mindless puppets or robot-like creatures."[232]

Similar discussions occur in the context of corporate law. A series of cases involved the standing and knowledge of companies that were unwittingly part of a Ponzi scheme. These "captive corporations" were seen as but the "robotic tools" of the scheme's architect.[233] Neither the corporations nor their investors could be imputed with the knowledge of the scheme. However, once released from the control of the Ponzi scheme,

---

225 *Id.* at 635.
226 *Id.*
227 *Id.*
228 *Id.*
229 *Id.* at 635 ("Neither does the evidence show negligence on the part of the defendant as a matter of law (if it shows negligence at all, which we need not decide).").
230 762 46 Cal. 3d 1092 (1988).
231 *Id.* at 1101.
232 *Id.* at 1131 (Anderson, J., concurring in part and dissenting in part.) The dissent cited heavily a 1983 article on whether religious converts were the "robots" of their church, entitled *Of Robots, Persons, and the Protection of Religious Belief* by Robert Shapiro, 56 SO. CAL. 1277 (1983).
233 Janvey v. Democratic Senatorial Campaign Committee, Inc., 712 F.3d 185, 190-92 (5th Cir. 2013).

these companies regained their status as separate corporate entities with standing to sue the architect for fraud and other damages.[234]

Conversely, under the "alter ego" theory, a corporation that is merely a defendant's robot is not entitled to treatment as a separate entity.[235] The doctrine says litigants can reach beyond a corporation to the personal assets of a company principal to the extent he or she uses the company to further purely personal interests.[236] But the standard is a high one. The plaintiff must establish that the controlled corporation acted "robot-like" and in "mechanical response" to the controller's "pressure on its buttons."[237] Only then will the court pierce the corporate veil on the alter ego theory.

The idea is that a robot is what a person or entity becomes when completely controlled by another. Such a person or entity is not capable of fault or knowledge, leaving the person behind the machine—the programmer—at fault. The effect is, interestingly enough, temporary. Thus, presumably the defendant who later learns to drive will be responsible for any accident she causes;[238] the victims of a religious cult may be deprogrammed and bring suit;[239] and an entity freed from the robotic control of a Ponzi scheme regains the usual rights of a corporation.[240] While a robot, however, no one sees, hears, or does evil.

C. THE MEANING OF ROBOT CASE LAW

To summarize the argument thus far: robots have played a role in American society for decades, occasionally resulting in interesting legal disputes. These include questions of how and when a robot can be said to represent a person, whether a robot extends people in ways the laws care about, and the conditions under which a person will be responsible for a

---

234 *Id. See also* Scholes v. Lehmann, 56 F.3d 750 (7th Cir.1995), *cert. denied sub nom.* African Enter., Inc. v. Scholes, 516 U.S. 1028 (1995)/.

235 *See* Partners Coffee v. Oceana Services and Products, 700 F. Supp. 2d 720, 737 (W.D. Pa. 2010). The alter ego theory should not be confused with the robot Walter Ego from Part I. But it is interesting to see the common verbiage.

236 *Id.*

237 Culbreth v. Amosa (Pty) Ltd., 898 F.2d 13, 15 (3d Cir. 1990).

238 *Id.*

239 *Id.*

240 *Id.*

robot's actions. Not only have robots been at the center of disputes, they have served from their earliest contact with American society as a judicial trope. Judges invoke robots over the decades to describe the limitations and advantages of human beings and to extend and limit a defendant's responsibility for misconduct. The way judges use robots to reason or frame issues may bear on the metaphors and frames they employ to decide cases involving robots and related technology.

In the remainder of the article, I offer some preliminary conclusions regarding the role of robots in American law. The first set involves the direction of the burgeoning field of robotics law and policy. Do past decisions shed light on questions scholars have already explored, such as liability, legal metaphor, and the interplay with race or gender? And do past decisions generate new questions that the field had not considered, such as whether certain legal categories require much greater nuance?

The second set involves the direction of the law itself. Judges appear to hold a very specific mental model of robots as discretionless machines. The model is outdated, if it ever held. This has led to tensions in the case law to date that will only become more acute as sophisticated robots become mainstream. At the same time, a deepening of familiarity with robotics, as well as a diversification in the sources of robot law, may signal a shift toward wiser policy going forward.

I. Robotics Law: An Early Agenda

Legal scholars have been writing about robotics here and there for some time.[241] The recent community around robotics law is especially robust, with a steady drumbeat of new research.[242] Collectively the work spans many different areas. But it tends to take one of several forms. The first is a sustained examination of a particular technology or legal issue. This

---

241 *See, e.g.*, Samuel N. Lehman-Wilzig, *Frankenstein Unbound: Towards A Legal Definition of Artificial Intelligence*, 13 FUTURES (1981). The same is true of other fields. For an early example in economics, see Herbert A. Simon, The Shape of Automation for Men and Management (1965).

242 This is in part due to the annual robotics law and policy conference colloquially known as "We Robot." The conference is in its fifth year, having been held at Stanford Law School, University of Washington, and twice at the University of Miami. The next We Robot will take place at Yale Law School in the spring of 2017.

work might address one or more dimensions of a single technology such as drones, driverless, cars, or surgical robots.[243] Alternatively, it may address the intersection of robotics with one or more of the following issues: (i) civil or criminal liability,[244] (ii) legal personhood or agency,[245] (iii) enforcement,[246] (iv) speech and intellectual property,[247] (v) race and gen-

---

243 *E.g.*, Gregory S. McNeal, *Targeted Killing and Accountability*, 102 GEO. L. REV. 681(2014) (drones); Ryan Calo, *The Drone as Privacy Catalyst*, 64 STAN. L. REV. ONLINE 29 (2014); Bryant Walker Smith, *Automated Vehicles Are Probably Legal in the United States*, 1 TEX. A&M L. REV. 411 (2104) (driverless cars); Jack Boeglin, *The Costs of Self-Driving Cars: Reconciling Freedom and Privacy With Tort Liability in Autonomous Drone Regulation*, 17 YALE J. L. & TECH. 171 (2015) (same); Julie Goodrich, *Driving Miss Daisy: An Autonomous Chauffeur System,* 51 HOUS. L. REV. 265 (2013) (same).

244 *E.g.*, *See also* PETER M. ASARO, A BODY TO KICK BUT NO SOUL TO DAMN: LEGAL PERSPECTIVES ON ROBOTICS, IN ROBOT ETHICS: THE ETHICAL AND SOCIAL IMPLICATIONS OF ROBOTICS (Patrick Lin et al., eds. 2012); Vladeck, *supra* note 163; Hubbard, *supra* note 163.

245 *E.g.*, Lawrence B. Solum, *Legal Personhood for Artificial Intelligence*, 70 N.C. L. REV. 1231 (1992); Christopher Stone, *Should Trees Have Standing? Revisited: How Far Will Law and Morals Reach? A Pluralist Perspective*, 59 S. CAL. L. REV. 1, 14 (1985) (discussing whether a robot could have standing); Steven Goldberg and Nancy Cruzan, *The Changing Face of Death: Computers, Consciousness*, 43 STAN. L. REV. 659 (1991); F. Patrick Hubbard, *'Do Androids Dream?': Personhood and Intelligence Artifacts*, 83 TEMP. L. REV. 405 (2011); SAMIR CHOPRA & LAURENCE F. WHITE, A LEGAL THEORY FOR AUTONOMOUS AGENTS (2011).

246 *E.g.*, Lisa A. Shay et al., *Confronting Automated Law Enforcement*, in ROBOT LAW (Ryan Calo, A. Michael Froomkin, and Ian Kerr, eds., 2016); Elizabeth E. Joh, Discretionless Policing: Technology and the Fourth Amendment, 95 CALIF. L. REV. 199 (2007).

247 For discussion of intellectual property, see, e.g., Annemarie Bridy, *Coding Creativity: Copyright and the Artificially Intelligent Author*, 2012 STAN. TECH. L. REV. 5, 21 (2012); Ralph D. Clifford, *Intellectual Property in the Era of the Creative Computer Program: Will the True Creator Please Stand Up?*, 71 TUL. L. REV. 1675, 1696-97 (1997) (arguing that patent law implicitly assumed a human inventor); Pamela Samuelson, *Allocating Ownership Rights in Computer-Generated Works*, 1185 U. PITT. L. REV. 47 (1985). For discussion of free speech, *see, e.g.*, Tim Wu, *Machine Speech*, 161 U. PENN. L. REV. 1495, 1496 (2013) ("The question of 'rights for robots,' if once limited to science fiction, has now entered the public debate."); Stuart Minor Benjamin, *Algorithms and Speech*, 161 U. PENN. L. REV. 1445 (2013).

der,[248] and (iv) privacy.[249] Technology or issue specific research is increasingly careful and rigorous and is already adding clarity and shape to the discipline. A limitation of the approach, however, is that it does not necessarily tell us much about the robot of robots *in general* across the law. The work it does is more specific.

Other research takes a step back and looks at the larger context of robotics law and policy. The advantage, arguably, of treating robotics as a broader phenomenon is that one sees connections across robots and legal domains. For example, the degree of control people exert over a robot could come up as an issue as easily in tort law (vicarious liability, res ipsa loquitur) as in the international law of the sea.[250] The issue with an approach that is less specific is that it also tends to be more speculative. The consensus among commentators is that robots will, someday, create legal and policy puzzles of a certain kind.[251] But we do not necessarily know in advance what they will be and, accordingly, we will be hard-pressed to identify models or solutions.

Consider again the conversation from the Introduction. Jack Balkin argues we cannot know how courts and others view robots.[252] Hence, we

---

248 *E.g.*, Sinziana M. Gutiu, *The roboticization of consen*t, in ROBOT LAW (Ryan Calo, A. Michael Froomkin, and Ian Kerr, eds., 2016); Ann Bartow, *Robots as Labor Creating Devices: Robotic Technologies and the Expansion of the Second Shift*, Proc. We Robot 2014, Apr. 4, 2014; Peter A. Asaro, *Will Black Lives Matter to Robocop?*, Proc. We Robot 2016, Apr. 1, 2016.

249 *E.g.*, Margot E. Kaminski, *Robots in the Home: What Will We Have Agreed To?,* 51 IDAHO L. REV. 661 (2015); Ryan Calo, *Robots and Privacy, in Robot Ethics: The Ethical And Social Implications of Robotics*, 195 (Patrick Lin et al., eds. 2012).
In addition, scholars work in related areas such as the effect of automation on the legal profession and the prospect of accountability and impartiality in the design and implementation of algorithms. *E.g.*, Daniel Martin Katz, *Quantitative Legal Prediction—Or—How I Learned to Stop Worrying and Start Preparing for the Data Driven Future of the Legal Services Industry*, 62 EMORY L.J. 909 (2013); FRANK PASQUALE, THE BLACK BOX SOCIETY: THE SECRET ALGORITHMS THAT CONTROL MONEY AND INFORMATION (2015); Solon Barocas and Andrew D. Selbst, *Big Data's Disparate Impact*, 104 CALIF. L. REV. (forthcoming 2016).

250 Calo, *supra* note 8, at 543-44 (discussing the concept of exclusive control in these contexts).

251 *See, e.g.*, Lehman-Wilzig, *supra* note 241. *See also supra*, note 245.

252 Balkin, *supra* note 12, at 45.

must wait and see to get a sense of what issues they will raise.[253] There is truth to Professor Balkin's claim: robotics is certainly advancing at a rapid pace and more and more people are coming into contact with robots outside of the contexts of manufacturing or warfare. The results for law and society are difficult to anticipate. What this article shows, however, is that robots have already raised specific and interesting issues that current technological trends stand to enhance. Focusing on these issues not only helps us understand how the law will come to mediate the mainstreaming of robotics, but it also suggests or grounds questions for the burgeoning field.

One example is the role of metaphor in judicial reasoning, already raised in the lead in to Part II. We sense that the metaphor or analogy a court or policymaker chooses to adopt around a new technology can influence its fate.[254] Thus, for example, thinking of email as analogous to a postcard could lead to less Fourth Amendment protection than thinking of email as analogous to a letter.[255] Believing encryption to be a form of speech could lead to First Amendment pushback should government seek to regulate encryption.[256]

Students of robotics law have not missed the importance of metaphor. Law professor Neil Richards and roboticist William Smart, for example, expressly call attention to the importance of metaphor and legal analogy in their work *How Should the Law Think About Robots?*[257] These authors conclude that courts should be careful to characterize robots as tools, albeit programmable ones, because doing otherwise runs the risk of committing what the authors call the Android Fallacy.[258] Two strengths of Richards and Smart's work are their recognition that metaphor matters and

253 *Id.*

254 *See supra*, notes __ to __ and accompanying text.

255 The Department of Justice has made this argument with mixed success. *See* SUSAN W. BRENNER, CYBERCRIME AND THE LAW: CHALLENGES, ISSUES, AND OUTCOMES 158 (2012).

256 Froomkin, *supra* note 193, at 879.

257 Neil Richards and William Smart, *How Should the Law Think About Robots?*, in ROBOT LAW 3 (Ryan Calo, A. Michael Froomkin, and Ian Kerr, eds., 2016). *See also* Meg Leta Jones and Jason Millar, *Hacking Analogies in the Regulation of Robotics*, in OXFORD HANDBOOK OF THE LAW AND REGULATION OF TECHNOLOGY (Karen Yeung, Roger Brownsword, and Eloise Scotford, ed. 2016).

258 Richards and Smart, *supra* note 257, at 4 ("Finally, we argue that one particularly seductive metaphor for robots should be rejected at all costs: the idea that robots are 'just like people'… We call this idea 'the Android Fallacy.'").

their sober-eyed examination of the present state of the technology. A weakness is their failure to imagine the ways robots do and will depart from the mental model they urge.

The case studies in Parts I and III may help us predict the sorts of metaphors judges are likely to select. More specifically, the case studies suggest that judges already think of robots as no more than programmable (or teleoperated) tools.[259] This is not necessarily ideal in all circumstances. For example, while it may be appropriate to hold the pilot accountable for failing to supervise the rudimentary robotic pilot of *Brouse v. United States*,[260] perhaps judges should resist the inclination to attribute liability to a person whenever he or she happens to be in the loop. We may not wish to incentivize the preservation of human control, even where less safe or efficient, merely to furnish a human "crumple zone" for liability.[261] I am aware that the liability still winds up landing on one or more humans — perhaps the manufacturer of the driverless car instead of whoever happens to be behind the skeuomorphic wheel. Even so, the metaphors and analogies we use influence *which* human pays the price for a robotic harm.

The case studies may inform other areas of scholarship as well. Consider the small but powerful literature surrounding the role of robotics in race and gender law.[262] Just as Jeannie Suk asks whether privacy is a woman,[263] robotics law and policy scholars may ask whether *Frye v. Baskin*—the case of the robotically driven Ford—comes out the same way if the gender of the litigants were reversed.[264] Would a court, particularly in 1950, find that a boy who could not drive was the robot of the girl who could? Scholars could also explore why the Ninth Circuit's example of a robot mostly clearly representing a person involves a robot Michael Jordan,[265] or whether there is anything to Judge Higginbotham's contention that black judges are expected to be robots where white judges are not,[266]

---

259 *See supra* Parts I and II.

260 83 F. Supp. 373, 374.

261 *See* Madeleine Elish, *Moral Crumple Zones: Cautionary Tales in Human Robot Interaction*, Proc. We Robot 2016, Apr. 1, 2016 (exploring the potential that people will be kept in otherwise fully autonomous systems for the sole purpose of absorbing liability).

262 *See supra*, note 248 (listing examples).

263 Jeannie Suk, *Is Privacy A Woman?*, 97 GEO. L. REV. 486 (2009).

264 S.W. 2d 630.

265 White, 971 F.2d 1395, 1399.

266 Local Union 542, Int. Union of Op. Eng., et al., 388 F. Supp. 155, 180.

and why it is that the concern over robotic testimony appears first and most often in the context of immigration.[267]

The preceding case studies not only illuminate existing debates, but hint at entirely novel questions scholars of robotics law and policy might explore. The literature speculates considerably as to who or what is to blame when a robot causes harm. It also grapples with whether the law should recognize legal, market, or intellectual behaviors initiated or executed by robots. Papers in the first category might ask whether the passenger or the manufacturer is responsible for a car accident in a driverless car.[268] Papers in the latter might investigate whether an artificial intelligence can act as a trustee, make a binding contract, or "create" an original work.[269]

There is next to no scholarship, however, tackling several of the very questions with which the courts already struggle. These tend to involve deep questions around the differences between people and machines. One question that remains untheorized in the early literature, for instance, is whether and to what extent a robot can represent a person. When you take or draw a picture of a person, it represents her.[270] But when you build a robot version of a person, does it represent the person in the same way? The Ninth Circuit at best equivocates, finding robots to represent people for purposes of common but not statutory law, though identically phrased.[271]

There is a similar struggle in the cases attempting to characterize robots for purposes of trade law. The court in *Louis Marx & Co.*, examining the tariff schedule for a windup toy, itself winds up in a dubious position that would likely benefit from scholarly engagement within and beyond the legal academy.[272] The court acknowledges that robots are machines that represent men—it consults a dictionary that says so—but nevertheless concludes that robot toys only represent robots, which are just machines.[273] There is even some tension between the appropriation cases and the cases involving tariffs. In the former, human features mean a robot can come to

267 *See supra*, notes 217-21 and accompanying text (listing cases).

268 *See supra*, note __.

269 *E.g.*, Solum, *supra*, note 245; Ian Kerr, *Ensuring the Success of Contract Formation in Agent-Mediated Electronic Commerce*, 1 ELECTRONIC COMM. R.J. 183-202 (2001); Bridy, *supra* note 247.

270 [cite].

271 *See supra*, 39-42 and accompanying text (citing White, 971 F.2d at 1395).

272 40 Cust. Ct. 610.

273 *Id.*

represent a person.[274] But in the latter, human features like those of the astronaut in *Lewis Galoob Co.* do not necessarily detract from its characterization of a robot.[275]

Another puzzle involves whether and under what circumstances a person operating a robot can be said to exist in the places to which the robot travels. I have highlighted two case studies—one involving claims to a particular right at sea, the other involving responsibility for the crime of entering a dwelling with intent to steal.[276] Thus, a court sitting in maritime generated a new category of possession to accommodate the changing nature of undersea exploration.[277] And a criminal court found that while a person can burglarize a bank with a metal robot, he cannot do so with a paper check.[278]

But there are myriad areas of law where this question could arise. One such area is space law. Recently the United States, with the passage of the US Commercial Space Launch Competitiveness Act of 2016, adopted the position that American individuals or companies who extract resources from an asteroid or the moon are free to alienate those resources.[279] Presumably two or more teams might compete for the right to mine the same extraterrestrial resource. Another area is remote or autonomous robotic surgery, which could raise questions of whether and where an operator or programmer must be board certified in order to perform a procedure in any given state.[280]

The issues presented are many and interesting. Do we allow persons to use a robot to extend themselves to new places only when a person cannot safely enter the environment? *Columbus America*, the case announcing the

---

274 *See* Wendt, 125 F.3d at 809 (permitting the lawsuit to go forward under California law because, unlike in White, the robots had molded skin like actual people).

275 *See* Lewis Galoob Co., 66 Cust. Ct. at 486 ("[T]he presence of a human face in an article which is otherwise incapable of representing any living being cannot make 'animate' what is, in totality, incapable of animation or life.").

276 *See* Parts I.D and I.E.

277 *See* Columbus-America Discovery Group, Inc, 742 F.Supp. at 1328-29.

278 *See* Davis, 958 P.2d 1083.

279 H.R.2262, Public Law No: 114-90 (2015-2016).

280 The first issue is similar to the kinds of problems of jurisdiction that the Internet created. If a surgeon in Maryland is operating on a patient in Ohio, must she be certified in Maryland, Ohio, or both? The second issue—presented, for instance, by Google's partnership with Johnson & Johnson to develop autonomous surgery robots—is whether the robot would need to be certified at all, let alone where.

doctrine of tele-possession, leaves this question open by limiting itself to the dangerous context of the high seas.[281] How much control must a robot afford over an environment for the person operating the robot to legally (or illegally) enter the space? What if there is a time delay between the command and its execution? What if the actions constitute a blend of human and machine control, or the work of many operators together as the case with military drones?[282]

Answering these and related questions in depth is outside the scope of this article. Nevertheless, the courts have already had to grapple with some of the difficult questions robots tend to spur, questions altogether outside of the existing legal literature. If these past cases are prologue, similar issues are on the horizon. And the robotics law and policy community is well-positioned to begin to lend analytic rigor to the debates.

II. Law and Contemporary Robotics

The bulk of this article is devoted to showing how courts have already come to grapple with robots in various ways. Robotics today is in the midst of a sea change, advancing at a breakneck pace. The field is headed toward a Cambrian explosion, referring to the great speed of the field's evolution and diversification.[283] What can the way courts have talked about robots in the past tell us about their reactions to this old but newly transformative technology? And how are we to proceed if we are to fashion a wise robotics law and policy going forward?

A path toward understanding how contemporary robots will interact with the law involves grappling with the differences and similarities between robotics today and the robots the law has already met. Such an exercise is crucial, both because the introduction of vastly more robots into society will amplify the effect of judicial mistakes regarding the technology, and because the technology itself has evolved past what the doctrine contemplates.

---

281 Columbus-America Discovery Group, Inc, 742 F.Supp. at 1328-29.

282 For a discussion of blended control, see Meg Leta Ambrose, *The Law and The Loop*, Proc. IEEE Ethics (2014).

283 Gill A. Pratt, *Is a Cambrian Explosion Coming for Robotics?*, 29 J. ECON. PERSP., 51 (2015).

Elsewhere I explore what I consider to be the three main qualities of robotics that pose particularly interesting legal and policy challenges.[284] The preceding case studies support my older arguments to a degree. For instance, I argue that the unique social valence of robots, i.e., the ways we are hardwired to react to a robot as though it were a social being, will pose novel challenges for law and policy.[285] And courts have indeed struggled with the ways robots might be surrogates for people beyond what one might expect with another technology. I also argue in past work that the physical embodiment of software code will play an important role in determining liability.[286] This claim sees modest support in the greater willingness of courts to assign blame to a person when that person relies on a robot in a way that leads to physical harm.

What I found most striking in my review of the case law, however, is how uniformly courts reject the prospect of emergent robot behavior. Emergence refers to the ability or tendency of a system to behave in complex, unanticipated ways.[287] Emergence has long been a gold standard in robotics.[288] The idea is to create robots that do not need to be programmed to solve tasks, at least not in the same way as an entirely bounded system. They can learn from experience and solve problems in ways their creators never envisioned.

This capacity for surprise, of course, is double-edged. It can result in a more efficient warehouse,[289] a new invention,[290] or an unanticipated style

284 Calo, *supra* note 8.

285 *Id.* at 545-49. *See also* Calo, *supra* note 249 (discussing social valence in the context of privacy).

286 Calo, *supra* note 8, at 532-37. *See also* Calo, *Open Robotics*, *supra* note 167.

287 Calo, *supra* note 8, at 539 (citing STEVEN JOHNSON, EMERGENCE: THE CONNECTED LIVES OF ANTS, BRAINS, CITIES, AND SOFTWARE 18-19 (2001)).

288 *Id.* at 538-39.

289 *See supra* notes 5 and 86 (discussing Amazon's use of robots).

290 ROBERT PLOTKIN, THE GENIE IN THE MACHINE: HOW COMPUTER-AUTOMATED INVENTING IS REVOLUTIONIZING LAW AND BUSINESS 51-52 (2009) (describing the "Creative Machine" that the company Gillette used to redesign its toothbrush). IBM's Watson, an artificial intelligence engine that beat the top Jeopardy players in an exhibition match, also generates new food recipes. The Bengali Butternut BBQ Sauce is pretty good.

of game play.[291] But it can also result in making people feel threatened or under assault. For example, police in Amsterdam investigated the designer of a Twitter bot—an autonomous software agent on the Internet—when it appeared to threaten harm to a local fashion show.[292] The bot threat had a real world impact through the show's concern about potential violence and the costs of a police investigation. Where a system is embodied, i.e., has the capacity physically to affect the world in itself, emergent behavior can result in actual bodily harm. Even absent embodiment an emergent system can threaten critical aspects of society, as when high-speed trading algorithms destabilize the stock market.[293]

From my study of how courts talk about robots, both in deciding cases about robots and in invoking the technology in the course of judicial reasoning, judges have a specific model of robots in mind. In American law to date, the robot is by definition a programmable machine.[294] It does exactly what its programmer tells it to do. It follows that a robot cannot be spontaneous enough to perform in a restaurant or responsible for the collision between two airplanes.[295] And it follows further that a person who acts like or is similar to a robot is not responsible for his or her own actions.[296]

The idea that a robot can only follow its programming is, on one level, correct. Robots do not somehow "decide" how to act, and they are not capable of deviating from the code that constitutes them. But judicial understandings of robots are also fundamentally wrong in crucial respects. Even if a robot will always run its code, that does not mean that a programmer anticipates, let alone intends, every action the robot takes. People can and do program robots to interact dynamically with their environment and to generate and attempt new solutions.[297] Part of the very appeal of these sys-

291 Deep Mind, an artificial intelligence company owned by Google, has taught an algorithm to play multiple Atari games. Often the style of play differs from that of the best gamers, even as it breaks all high scores.

292 *See* Kashmir Hill, *Who do we blame when a robot threatens to kill people?*, FUSION, Feb. 15, 2015, http://fusion.net/story/48656/who-do-we-blame-when-robots-threaten-to-kill-people/.

293 *See* Calo, *supra* note 8, at 541 (citing Tom C.W. Lin, *The New Investor*, 60 UCLA L. REV. 678 (2013)).

294 This definition is still popular in certain disciplines such as electrical engineering.

295 *See supra* Parts I.B and I.E.

296 *See supra* Part II.C.

297 *See* Calo, *supra* note 8, at 538.

tems is that they come up with novel strategies and results by approaching problems in a way no human would.[298]

Courts have yet to come into significant contact with emergent systems. When they do, judges will have to shed their current conceptions of the technology or they will make analytic mistakes. Thus, while the research underpinning this article corroborates several of my earlier hypotheses, it amends others. I have assumed courts confronted with emergent behavior will recognize that its creators did not foresee or intend the behavior.[299] The logical step would be the refusal to find the necessary mens rea in criminal law or foreseeability in tort, leading to would-be prosecutors with no defendants and victims without redress.[300]

My recent research convinces me that courts may assume this problem away for a time. The idea that robots do the specific bidding of people appears firmly ingrained in the judicial imagination. Confronted with an emergent mishap, courts may lay blame with the creator or operator no matter what arguments these defendants deploy. There are, perhaps, advantages to this approach. But the disadvantages are also significant. As the Supreme Court recently reaffirmed, the Constitution frowns on using even negligence as the intent standard for crimes.[301] And the approach in tort would go beyond even strict liability, where courts dispense with an inquiry into fault but nevertheless require foreseeability to satisfy proximate causation.[302]

The preceding raises an interesting institutional question: who should decide how law evolves with robotics? Another difference today, beyond advances in technology, is that officials other than the judiciary have become primary sources of robotics policy. Increasingly, legislatures and agencies are passing rules about robots. A full accounting for these laws is beyond the scope of this particular article. But just a few examples include dozens of state laws regulating driverless cars, the domestic use of drones, and insurance for telemedicine.[303]

---

298 *Id.*

299 *Id.* at 541-45, 554-55.

300 *Id.* at 554-55.

301 Elonis v. United States, 575 US. __, at *13-14 (2014) (rejecting a negligence standard in a criminal case).

302 Calo, *supra* note 8, at 554-55.

303 *E.g.*, S.B. 1298, 2012 Leg., Reg. Sess. (Cal. 2012) (authorizing autonomous vehicles); A.B. 511, 2011 Leg., 76th Sess. (Nev. 2011) (same); S.B. 313, 2013 Leg., 77th Sess. (Nev. 2013) (regulating autonomous vehicles); S.B. 1134, 62nd Leg.,

These laws are not necessarily wise or perfectly informed. The state of Nevada had to repeal and rewrite its pioneering driverless car legislation within a year because of an unanticipated impact on automakers.[304] But the new laws demonstrate that the courts are not the only place where robotics law and policy is made. A plausible scenario suggests that federal and state lawmakers wind up setting the new tone, tenor, and content of robotics law. Recently, the National Highway Transportation Safety Administration responded to Google's request to clarify whether driverless car software could quality as a "driver." The NHTSA responded that, under certain circumstances, software could be a driver—an opinion likely to influence other judicial, regulatory, and legislative bodies down the line.

New legal institutions could play a role. Agencies famously serve as repositories of expertise.[305] I and others argue in favor of a new technology commission, a kind of NASA-for-everything that can act as a repository of knowledge about robots to guide legal actors, including courts.[306] This article focuses on case law because state and federal courts have been the main points of contact between robots and the law in the past half-century. New configurations are always possible.

The mental models of judges may also change with time. Robotics experts may come before the court; technically minded organizations and individuals may file amicus briefs. In several of the cases in Part I, we see the court turning to the dictionary.[307] This is interesting insofar as the definition of a robot changes with time. Thus, imagine if the court in [insert case] were to apply the contemporary definition of a robot as a "machine that looks like a human being and performs various complex acts (as walking or talking) of a human being" to the toy under examination.[308]

1st Reg. Sess. (Idaho 2013) (placing limits on domestic use of drones); S.B. 1587, 98th Gen. Assemb., Reg. Sess. (Ill. 2013) (same); Va. Code § 38.2-3418.16 (requiring insurance to cover telemedicine).

304 NRS § 482A.020 (repealed 2013); NRS § 482A.30 (2011).

305 *See* Chevron, U.S.A., Inc. v. Natural Res. Def. Council, Inc. 467 U.S. 837, 866 (1984).

306 *See* Ryan Calo, *The Case for a Federal Robotics Commission* (2014), http://www.brookings.edu/research/reports2/2014/09/case-for-federal-roboticscommission; Bruce Schneier, *The Internet of Things Will Be the World's Biggest Robot*, FORBES (Feb. 2, 2016), http://www.forbes.com/sites/bruceschneier/2016/02/02/the-internet-of-things-will-be-the-worlds-biggest-robot/#27d41a853162.

307 *See supra*.

308 "Robot." *Merriam-Webster.com. Merriam-Webster*, n.d. Web. 10 Feb. 2016.

But as we have also seen, courts are perfectly capable of ignoring definitional elements that go against their intuitions about the technology.[309]

The broadening of sources of robotics law may prove a good or bad thing.[310] Lawmakers and agency officials may be better positioned than courts to gain the expertise and appreciation of just how far robotics has come, or else introduce new and unwelcome complexities. New policy frameworks, coupled with good experts and amicus briefs, could, on a charitable reading, help update the mental models of courts and lead to analytically sound outcomes. Here I agree with Professor Balkin: we will only understand the true impact of robots on American law and legal institutions in the fullness of time.

CONCLUSION

This article offered nine new case studies to illuminate the role of robots in American law. The first set considered robots as *objects* of American law, i.e., as artifacts in the world that have occasioned legal disputes. The second set considered robots as *subjects* of the judicial imagination, i.e., as metaphors or similes that support a particular verdict. Few of the decisions that make up these studies appear in the legal literature to date—even within the burgeoning field of robotics law.

Examining these cases together yields a series of valuable insights. One is that robots tend to blur the line between person and instrument. A robot is an artifact, but one holding special fascination and significance. Courts have struggled with the status of robots, asking how robots can be said to represent, imitate, extend, or absolve people. Judges also invoke robots in a variety of contexts when a person is acting, or being asked to act, outside of the typical bounds of a human being.

A second is that judges may have a problematically narrow conception of what a robot is. There are, unsurprisingly, conflicts and tensions in early robot law. Courts are ambivalent about each of the questions they confront and results vary with context. But there seems to be an odd consensus with

309 *See supra.*

310 For a recent argument that courts are surprisingly well-positioned to cultivate expertise, see Elizabeth Fisher et al., *Rethinking Judicial Review of Expert Agencies*, 93 TEXAS L. REV. 1681 (2015).

respect to the judge's mental model of a robot: it is a programmable machine, by definition incapable of spontaneity.

If this definition or robots were ever true, it is not true today. Contemporary robots range in sophistication, with some systems solving and creating problems in ways neither the programmer nor the public would have anticipated. The mismatch between what a robot is and how courts are likely to think of robots will only grow in salience and import over the coming decade.

The story of robots and the law is only beginning. Robots are rapidly entering the mainstream and are likely to alter the legal landscape in ways prosaic and profound. But we should not assume we write on a clean slate. Robots have been a part of American society for half a century. And as the preceding pages show, they have already generated a limited but significant body of law that awaits exploration.

Technology Law in China: Computer Hacking, Cloud Computing and Robotics as Examples

Xiaoyan Huang,
University of Würzburg

A. Introduction: The Relationship between Technology and Law

In his book *High-tech Society: Story of the Information Technology Revolution*, Tom Forester wrote: "The nations of the world are caught up in a revolution: a technological revolution, which is bringing about dramatic changes in the way we live and work – and maybe even think".[1] Technological development is moving faster and faster now. Industry must learn to adapt to the pace of this technological development. Life on earth will be fundamentally changed by the onslaught of technology. Developments in technology, and new ideas from high tech, will exert a huge influence over our societies. They will also raise new legal issues, which, over the course of time, will become areas in need of state regulation. That is why the question of what the proper relationship should be between law and technology has become so important.

The relationship between law and technology is characterized by reciprocal effects influencing both, and by the respectively high importance of each for society.[2] The law is a social factor and in modern society it is influenced by technology. Developments in technology lead to changes both in the substantive law and in the procedural law. In the opposite direction, the law organizes and steers the development of technology.[3] It guarantees that technological progress and development are informed and guided both by the provisions and values of our democratic normative order and by the rules of modern science. Moreover, the law guarantees the freedom to conduct scientific research and protects intellectual property rights which flow from that research.

---

1 Forester, *High-tech Society: Story of the Information Technology Revolution*, 1987, p. 1.
2 Skistims, *Smart Homes*, 2016, p. 103.
3 Kloepfer, *Technik und Recht im wechselseitigen Werden*, 2002, p. 81.

B. A Newly Emerging Legal Field: Technology Law

The rapid development of technology is leading to a situation where the material civilization of humanity can hardly be separated from technology. The application of technology is more or less inseparable from the activities of everyday life. On the one hand, its application promotes the improvement of human well-being, but on the other hand, certain phenomena are arising as a consequence of the application of technology which have consequences for the entire society. To regulate this, new law, or respectively, new statute law will be necessary. This is why technology and law are being joined to create the new legal subject area called *technology law*.

In technology law, two independent scientific disciplines are seen to converge: engineering sciences and legal science.[4] That is why technology law is regarded as an interdisciplinary research area.[5] Technology law has the following functions: (1) to recognize, analyze and respond to the most important new legal issues, on the basis of applicable law; (2) to Fidentify and assess as early as possible legal barriers which might inhibit technological development or the use of new technologies; (3) to determine whether new laws or changes to existing norms are needed to deal with problems arising in the context of technological development.[6]

The National People's Congress of China (the national parliament of the PRC – abbreviated NPC) has placed great emphasis on legislation in the field of technology law over the last 20 years. Important statutes which have been adopted include: the Scientific and Technological Progress Law, the Scientific and Technological Popularization Law, and the Chinese Law on Promoting the Transformation of Scientific and Technological Achievements.

---

4 Ensthaler, Gesmann-Nuissl & Müller, *Technikrecht*, 2012, p. 1.

5 For fundamental research and interdisciplinarity in Germany, see Hilgendorf, "Zur Lage der Juristischen Grundlagenforschung in Deutschland heute" in Brugger, Neumann & Kirste, *Rechtsphilosophie im 21. Jahrhundert*, 2008, p. 121 et seq.

6 Hilgendorf, "Die strafrechtliche Regulierung des Internets als Aufgabe eines modernen Technikrechts", *Juristen Zeitung* 2012, p. 827.

C. Legal Challenges in Industrie 4.0 – Developing Legal Solutions Relevant to Made in China 2025?

Industrie 4.0[7] will involve all sorts of technological innovation. It is possible that the law will not be able to keep pace with these developments. But it is important to keep in mind that legal regulation plays an important role in promoting technological development. The new and innovative ways technology will develop under *Industrie 4.0* will certainly create challenges for the law. German legal experts point to tort liability, IT security or, respectively, data security, and data protection (i.e. privacy).[8] They are currently looking very closely at potential legal problems within the legal framework of the present law.

The legal system in Germany is famous around the world and is considered to be a model legal system within the civil law legal tradition. But why is it worthwhile for us to occupy ourselves with Chinese law? One way to answer this question would be to simply point out the incredible economic importance of the People's Republic of China. After the chaos of the Cultural Revolution in the 1966 – 1976 period, in which both the economy and the legal system were completely destroyed, in 1978 the new Chinese leadership seized the opportunity not only to open and reform the economic system, but also to gradually build a new legal system. During the last 30 years, China has made great progress both in the economic field and in the field of law. At the moment the Chinese government is planning the *Made in China 2025* project,[9] which will pursue aims similar to the German *Industrie 4.0* project. In addition, the State Council intro-

---

7 "Industrie 4.0" is a term that was introduced at the Hanover Fair in 2013 by the Federal Ministry of Education and Research and by the Federal Ministry of Economics and Technology. This term describes future industrial production in Germany. "Industrie 4.0" is understood as the fourth industrial revolution, oriented towards intelligent production. Previous industrial revolutions were characterized by the use of the steam engine (1st revolution), mass production together with assembly lines (2nd revolution), and electronics and digitization (3rd revolution), see Roth, "Industrie 4.0 – Hype oder Revolution?", *Einführung und Umsetzung von Industrie 4.0*, 2016, p. 5.

8 Cf. Hötitzsch, "Juristische Herausforderungen im Kontext von 'Industrie 4.0' – Benötigt die vierte industrielle Revolution einen neuen Rechtsrahmen?", in Hilgendorf & Hötitzsch (eds.), *Das Recht vor den Herausforderungen der modernen Technik*, 2015, p.80 *et seq.*

9 For more on this project, see http://english.gov.cn/2016special/madeinchina2025/; For additional detail on the relationship between "Industrie 4.0" and "Made in Chi-

duced a five-year plan on China's national informatization in December 2016. The plan includes the promotion of cutting-edge technologies such as cloud computing and the Internet of Things by granting further funding. The information infrastructure will also be extended in order to provide rural areas with cable Internet. It is expected that with further development in the area of technology, Chinese legal scientists will increasingly seek ways to harmonize the relationship between the law and technology. In light of the fact that the Chinese legal system as it stands now is less than fifty years old, Chinese lawyers will need to make use of the experience gained by their legal colleagues abroad in order to find the best solutions for China. In these endeavours, the Chinese are placing great focus on German developments in particular. This is because of the extremely high reputation which German law enjoys in China. In point of fact, the Chinese want to take German law as their model and then adapt it to Chinese conditions in order to improve the Chinese regulatory framework in the area of technology.

D. The Example of Computer Hacking and Criminal Law

I. Internet Crime and Computer Hacking

The Internet is one of the most important human inventions in the 20th century. One could say that the World Wide Web has truly become a kind of virtual reality. Looking back on human history, one can clearly see that the invention and dissemination of new technologies has always led to the transformation of the way people lived in society; life is experienced differently and new social norms arise. On the one hand, the development of Internet technology today is improving the quality of life, and, by promoting economic growth, it is allowing many people to realize a level of prosperity which previous generations could only dream of. On the other hand, concerns are also increasingly being raised about the potential negative impact Internet technology may entail. This technology harbours risks of

na 2025" see Wübbeke & Conrad, *Industrie 4.0: Deutsche Technologie für Chinas industrielle Aufholjagd?*, 2015, available at http://www.merics.org/fileadmin/templates/download/china-monitor/China_Monitor_No_23.pdf; *Made in China 2025: Die Kampfansage an Deutschland*, 2015, available at http://www.zeit.de/wirtschaft/2015-05/china-industrie-technologie-innovation.

social harm, which could be realized at any time, for example, in the perpetration of criminal offences. Thus, the spread of Internet technology has opened many doors for criminals to pursue their illicit goals. Hacking is a good example of this.

Russell and Cunningham defined hacking as criminal behavior in which a malicious intruder surreptitiously enters into a computer system to gain access to sensitive information.[10] Hacker crime has the following characteristics: firstly, exploitation of technology; secondly, the increased prevalence of crimes committed via computer technology; thirdly, many younger participants.[11] The best-known methods employed by hackers include Trojan horses, worms, viruses, phishing, key loggers, backdoors and exploits.

II. Criminal Law Provisions in China to Combat Hacking and Associated Problems

In 1997 the National People's Congress adopted the revised Criminal Law of the PRC (i.e. Chinese criminal code – abbreviated CLPRC) (中华人民共和国刑法), which due to its considerably expanded scope alone differs significantly from its predecessor from 1979. The provisions of the new criminal code relevant to the offence of hacking are § 285,[12] § 286[13] und

10 See Russell & Cunningham, *Das Hacker-Buch*, (tr. Travis), 2001, p. 25 et seq.

11 See *HACKER 'Mixter' mischt das Netz auf*, 2000, available at http://www.focus.de/politik/deutschland/hacker-mixter-mischt-das-netz-auf_aid_181759.html.

12 An English translation of the Chinese Criminal Code is available at http://www.lawinfochina.com.
Article 285 CLPRC (1997): Whoever violates state regulations and intrudes into computer systems with information concerning state affairs, construction of defense facilities, and sophisticated science and technology is be sentenced to not more than three years of fixed-term imprisonment or criminal detention.

13 Article 286 CLPRC (1997): Whoever violates states regulations and deletes, alters, adds, and interferes in computer information systems, causing abnormal operations of the systems and grave consequences, is to be sentenced to not more than five years of fixed-term imprisonment or criminal detention; when the consequences are particularly serious, the sentence is to be not less than five years of fixed-term imprisonment.
Whoever violates state regulations and deletes, alters, or adds the data or application programs installed in or processed and transmitted by the computer systems,

§ 287[14] CLPRC. Illegal entry into a computer system and the destruction of computer information systems are criminal offences, under § 285 and § 286 respectively. Conventional crimes committed by means of computer hacking are covered in § 287. The 7th Amendment of CLPRC, which was approved on 28 February 2009 by the 7th meeting of the Standing Committee of the 11th National People's Congress, introduced two new offences into § 285,[15] namely: (1) illegal provision of data from computer information systems, and illegal use of computer information systems, and; (2) the delivery of programs and resources for the illegal invasion and illegal use of computer information systems. The 9th amendment of CLPRC, which was approved on 8.29.2015 by the 16th meeting of the Standing Committee of the 12th National People's Congress, introduced several

---

and causes grave consequences, is to be punished according to the preceding paragraph.
Whoever deliberately creates and propagates computer virus and other programs which sabotage the normal operation of the computer system and cause grave consequences is to be punished according to the first paragraph.

14 Article 287 CLPRC: Whoever uses a computer for financial fraud, theft, corruption, misappropriation of public funds, stealing state secrets, or other crimes is to be convicted and punished according to relevant regulations of this law.

15 Article 285 CLPRC (2011): Whoever violates state regulations and intrudes into computer systems with information concerning state affairs, construction of defense facilities, and sophisticated science and technology is be sentenced to not more than three years of fixed-term imprisonment or criminal detention.
Whoever, in violation of the state provisions, intrudes into a computer information system other than that prescribed in the preceding paragraph or uses other technical means to obtain the data stored, processed or transmitted in the said computer information system or exercise illegal control over the said computer information system shall, if the circumstances are serious, be sentenced to fixedterm imprisonment not more than three years or criminal detention, and/or be fined; or if the circumstances are extremely serious, shall be sentenced to fixed-term imprisonment not less than three years but not more than seven years, and be fined.
Whoever knows that any other person is committing the criminal act of intruding into or illegally controlling a computer information system and still provides programs or tools for such a person shall, if the circumstances are serious, be punished under the preceding paragraph.

new provisions, namely § 286a,[16] § 287a[17] and § 287b[18] as well as criminalizing companies under § 285[19] and § 286.[20]

---

16 Article 286a CLPRC (2015): Any network service provider that fails to perform the information network security management obligation as prescribed in any law or administrative regulation and refuses to make corrections after being ordered by the regulatory authority to take correction measures shall be sentenced to imprisonment of not more than three years, criminal detention or surveillance in addition to a fine or be sentenced to a fine only under any of the following circumstances.

17 Article 287a CLPRC (2015): Whoever commits any of the following conducts by using the information network shall, if the circumstances are serious, be sentenced to imprisonment of not more than three years or criminal detention in addition to a fine or be sentenced to a fine only:
(1) Establishing a website or a communication group mainly for committing fraud, teaching on how to commit a crime, producing or selling any prohibited or controlled article, or committing any other illegal or criminal activity.
(2) Issuing any information on the production or sale of drugs, guns, obscene articles, or any other prohibited or controlled article or any other illegal or criminal conduct.
(3) Issuing any information for committing fraud or any other illegal or criminal activity.

18 Article 287b (2015) CLPRC: Whoever, while obviously aware that any other person is committing a crime by using an information network, provides Internet access, server custody, network storage, communication transmission or any other technical support, or provides advertising, payment settlement or any other assistance for the crime shall, if the circumstances are serious, be sentenced to imprisonment of not more than three years or criminal detention in addition to a fine or be sentenced to a fine only.
Where an entity commits any crime as provided for in the preceding paragraph, the entity shall be sentenced to a fine, and its directly responsible person in charge and other directly liable persons shall be punished in accordance with the provisions of paragraph 1.
Whoever commits any other crime while committing a crime as mentioned in the preceding two paragraphs shall be convicted and punished according to the provisions on the crime with the heavier penalty.".

19 Article 285 CLPRC (2015): Whoever violates state regulations and intrudes into computer systems with information concerning state affairs, construction of defense facilities, and sophisticated science and technology is be sentenced to not more than three years of fixed-term imprisonment or criminal detention.
Whoever, in violation of the state provisions, intrudes into a computer information system other than that prescribed in the preceding paragraph or uses other technical means to obtain the data stored, processed or transmitted in the said computer information system or exercise illegal control over the said computer information system shall, if the circumstances are serious, be sentenced to fixedterm imprisonment not more than three years or criminal detention, and/or be fined; or if the cir-

The criminal provisions relating to hacker crime are found in CLPRC Chapter 6 "endangering the administrative order of society". This means that the legal interest which is infringed by hacker crime is called "the administrative order of society". But it is apparent that hacker crime is very harmful to society so that one could say it even endangers public security. It the NPC the provisions on hacker crime were put into the wrong chapter of the statute (namely chapter VI "Crimes of Disrupting the Order of Social Administration").

CLPRC § 30 regulates the criminal liability of legal persons, i.e. companies and organisations, referring to them as "unit".[21] The 9th amendment of CLPRC extended criminal liability to legal persons under § 285 and § 286 CLPRC. It is not understandable why the NPC did not extend

---

cumstances are extremely serious, shall be sentenced to fixed-term imprisonment not less than three years but not more than seven years, and be fined.
Whoever knows that any other person is committing the criminal act of intruding into or illegally controlling a computer information system and still provides programs or tools for such a person shall, if the circumstances are serious, be punished under the preceding paragraph.
Where an entity commits any crime as provided for in the preceding three paragraphs, the entity shall be sentenced to a fine, and its directly responsible person in charge and other directly liable persons shall be punished according to the provisions of the applicable paragraph.

20 Article 286 CLPRC (2015): Whoever violates states regulations and deletes, alters, adds, and interferes in computer information systems, causing abnormal operations of the systems and grave consequences, is to be sentenced to not more than five years of fixed-term imprisonment or criminal detention; when the consequences are particularly serious, the sentence is to be not less than five years of fixed-term imprisonment.
Whoever violates state regulations and deletes, alters, or adds the data or application programs installed in or processed and transmitted by the computer systems, and causes grave consequences, is to be punished according to the preceding paragraph.
Whoever deliberately creates and propagates computer virus and other programs which sabotage the normal operation of the computer system and cause grave consequences is to be punished according to the first paragraph.
Where an entity commits any crime as provided for in the preceding three paragraphs, the entity shall be sentenced to a fine, and its directly responsible person in charge and other directly liable persons shall be punished according to the provisions of paragraph 1.

21 Article 30. CLPRC: A company, enterprise, institution, organization, or group which commits an act endangering society that is considered a crime under the law shall bear criminal responsibility.

criminal liability under § 287 to legal persons. Therefore under § 287 criminal liability for conventional crimes, committed by means of computers, may be incurred by natural persons but not by legal persons.

CLPRC § 285 subsection 2 covers the crimes of "illegal acquisition of data" and "illegal control in the computer information systems". But the scope of protection of the two offences according to the black letter text of CLPRC does not include state affairs, defence apparatus, sophisticated science or technology. This means that CLPLC § 285 subsection 1 contains two offences, namely: (1) hacking to obtain data, and (2) hacking to exercise control over a computer. The wording of these two paragraphs was obviously not drafted in a logical manner by the NPC.

III. Prospects for Future Criminal Law Provisions against Computer Hacking in China

The criminal activities of hackers are an enormous problem for all nations. In connection with "Industry 4.0" and "Made in China 2025", a uniform international system of judicial cooperation (law enforcement) must be established in order to fight effectively against this crime. It is essential that China be part of this system. China must first work more actively and intensively with the Council of Europe, the EU and Germany in this area and exchange experiences, by signing the Cybercrime Convention in the future,[22] taking as a role model the legislative experiences of the EU directive on attacks against information systems,[23] in order to achieve harmonization between domestic and international legislation in criminal law provisions to combat hacking.

The relevant provisions on hacker crime in the CLPRC focus much more on the protection of computer systems than on the protection of data. In connection with the information society, or, respectively, with *Made in China 2025*, the technology referred to as "big data" or "data lake" plays a decisive role. That is why the focus of the criminal law on the protection

22 Übereinkommen über Komputerkriminalität, 2001, SEV Nr. 185, available at http://www.cr-online.de/2001_11-23_Europarat_Uebereink._Computerkriminalitaet.pdf.

23 Directive 2013/40/EU of the European Parliament and of the Council of 12 August 2013, available at http://eur-lex.europa.eu/LexUriServ/LexUriServ.do?uri=OJ:L:2013:218:0008:0014:en:PDF.

of computer systems no longer fits the developmental requirements of Chinese industry. Ensuring data security should become the core of the new Chinese Internet criminal law during the *Made in China 2025* project. If that is so, offences like "data espionage", "interception of data" and "modification of data" should be covered in detail in CLPRC. The focus of protection of Chinese criminal law should therefore be shifted to the protection of data, rather than the protection of computer systems.

Hacker crime is a kind of computer crime or cybercrime. It is a category of technology crime. Hacker crime is not possible without the Internet. That means that this kind of crime is conducted via the Internet. A perpetrator needs a high level of technological skills to commit hacking offences. The Internet is simply a form of technology. In order to better prevent hacker criminality, the NPC must first understand how this technology offers criminals opportunities and space to commit crimes. After grasping how the Internet is designed and functions, it can set about analyzing hacker crime much more extensively and efficiently.

Preventing hacker crime will necessarily involve a more detailed examination of the technology. If one evaluates a new technology, the risk that it can be used to commit crimes should also be assessed. An assessment of whether a technology can be used to commit crimes should be a routine task when new technology is evaluated. If a new technology may promote or make easier certain forms of crime, i.e. giving criminals opportunities to commit crimes, then increased surveillance is necessary. Of course crime still has to be attributed to human beings. Human beings invent technology and it is closely associated with them. In the future the law should be amended to stipulate that further developments in technology should aim not only to reduce opportunities to commit crime, but also to increase penalties for perpetrators. In the future, the trend should be towards developing technology in order to make the prevention of crime easier.

Made in China 2025, together with the five-year plan on China's national informatization, will bring the development of Internet technology in China to a new and higher level. The future use of Internet technology will in any case surpass our imagination, and continue to change our way of working and living. The Chinese Criminal Code, as it stands today, will not be able to cope with the criminal issues raised by these new technological developments. In order to achieve a balance between technological development and the maintenance of law and order in society as we have come to expect it, a new technology criminal law will be required. The term *technology criminal law* could, on the one hand, be understood as a means for the protection of technology (as a kind of intellectual property) by the criminal law. One could say that the ownership of intellectual property in technology represents a legal interest which is worthy of protection by the general criminal law.[24] On the other hand, technology criminal law could be understood as the use of criminal law as a means of preventing the use of technology by human beings in order to commit crimes. The misuse of the technology entails the risk that the particular legal interests protected by the criminal law could be infringed.[25] This is why technology criminal law has two functions: to promote the development of technology and to control and sanction the misuse of technology.[26]

E. The Example of Cloud Computing and Data Protection Law

I. Data Related to Persons

The development of the information economy and the dissemination of information technology mean that data has become one of the most important factors in production. This type of economic development is gradually transforming the industrial economy, which depends on capital and energy, into the information economy, which is dependent on computerization and virtualization. During this transformation, various different kinds of data will become essential factors of economic life; above all, this includes data relating to persons or their preferences, which can be exploited commercially or otherwise. Person-related data are the most important kind of data.

Personal data have value both for public administration (i.e. government purposes) and for use in trade and commerce. Non-public institutions, including companies with profit as their motive, collect personal data as a commercial instrument to help them improve their bottom line. The law is not settled as to who owns these data. The analysis of these data is very useful for commercial purposes. By gathering personal data, companies can create better customer profiles, which facilitates decision mak-

24 Vassilaki, "Technikstrafrecht" in Schulte (ed.), *Handbuch des Technikrechts*, 2011, p. 385, 386.
25 *Ibid.*, p. 386.
26 *Ibid.*, p. 387.

ing. Good data also help companies to develop better market forecasts. For example, a bank can sell personal data about its customers to an insurance company, so that this company can direct a marketing campaign towards those customers. As can be seen, personal data are of considerable commercial value, which is why the private sector is eager to collect such data in order to use them for its own purposes. This increases the risk that data will be unlawfully sold or otherwise misused.

II. Cloud Computing and Personal Privacy Issues

Rapid economic development is leading companies to rapidly increase the quantities of data they collect. Modern enterprises are therefore facing the question of how to manage and analyze the copious data they collect. The traditional data collection and storage systems firms have used up to now are no longer able to cope with such large scale data processing tasks. That is why "cloud computing" is getting so much attention at present. It allows "supercomputing" (data analysis) and data storage to be conducted via the Internet at remote locations. In the last several years cloud computing has also become the computing model which the IT industry is focusing on. One would therefore have to say that cloud computing has become an inevitable trend in the future management of comprehensive data sets. In China there are more than 300 cloud providers, the most well-known of which worldwide include Alibaba Group,[27] Tencent,[28] Baidu[29] and Huawei.[30]

There is currently very intensive discussion both at EU level and in Germany on legal issues being raised by the cloud computing model related to the protection of personal data.[31] From the point of view of the Article 29 Data Protection Working Party, cloud computing involves two ma-

---

27 https://intl.aliyun.com/zh (in Chinese).
28 https://www.qcloud.com/ (in Chinese).
29 https://cloud.baidu.com/ (in Chinese).
30 http://e.huawei.com/cn/solutions/technical/cloud-computing (in Chinese).
31 See Heidrich & Wegener, "Sichere Datenwolken – Cloud Computing und Datenschutz" in *MultiMedia und Recht* 2010, p. 803; Folizick & Plankemann, "Cloud Computing and Compliance: Probleme und Lösungsansätze" in *Corporate Compliance Zeitschrift* 2015, p. 180; Lewinski & Hermann, "Cloud vs. Cloud – Datenschutz im Binnenmarkt", *Zeitschrift für Datenschutz* 2016, p. 467; Nägele & Jacobs, "Rechtsfragen des Cloud Computing", *Zeitschrift für Urheber- und Medien-*

jor privacy risks: on the one hand, the lack of control has been pointed out. Cloud users may no longer have exclusive control over their data and cannot take the technological and organizational measures necessary to ensure the availability, integrity, confidentiality, transparency, isolation, intervenability and portability of the data. On the other hand, lack of information on processing (transparency) is mentioned, and insufficient information on the processing processes of a cloud service pose a risk for both the data controller and the affected persons.[32]

Companies connected to *Made in China 2025* in China are faced with data protection challenges because of their use of cloud computing. What is at issue here is not the personal data of customers, but rather the data of employers. For these companies, the urgent task is to determine precisely the purposes for which technologies such as cloud computing are used and to restrict data transfer precisely to what is necessary on the basis of this determination. What is obvious, however, in respect of data protection law, is that the NPC has urgent work in front of it.

III. The Legal Situation and Problems of Data Protection Law in China

It is important to understand that currently, full legal protection of personal data does not exist in China. Indeed, up to now the NPC has still not passed independent legislation on the protection of personal data. On 23 April 2013 the Supreme People's Court, the Supreme Procuratorate and the Ministry of Public Security for the first time provided a definition of "personal data" in China in their "Notification on Legally Punishing Criminal Activities that Infringe upon Citizens' Personal Information".

The China Cybersecurity Law (CCL) was enacted by the NPC on 7 November 2016 at its 24th session. The relevant provisions of the act in

---

recht 2010, p. 281; Kroschwald, "Kollektive Verantwortung für den Datenschutz in der Cloud – Datenschutzrechtliche Folgen einer geteilten Verantwortlichkeit beim Cloud Computing", *Zeitschrift für Datenschutz* 2013, p. 388; Molnár-Gábor & Korbel, "Verarbeitung von Patientendaten in der Cloud – Die Freiheit translationaler Forschung und der Datenschutz in Europa", *Zeitschrift für Datenschutz* 2016, p. 274.

32 European Commission Article 29 Working Party, Opinion 05/2012 on Cloud Computing, 2012, pp. 6 – 7, available at http://ec.europa.eu/justice/data-protection/article-29/documentation/opinion-recommendation/files/2012/wp196_en.pdf.

respect of the protection of personal data are contained in Chapter 4 ("Network Information Security"), Articles 40 – 45.[33]

The NPC took advantage of the legislative experience accrued abroad, establishing in the act the following key principles:

---

33 The English translation of the "China Cybersecurity Law" is available at http://www.chinalawtranslate.com/cybersecuritylaw/?lang=en.
Article 40 CCL: Network operators shall strictly maintain the confidentiality of user information they collect, establish and complete user information protection systems.
Article 41 CCL: Network operators collecting and using personal information shall abide by the principles of legality, propriety and necessity; make public rules for collection and use, explicitly stating the purposes, means, and scope for collecting or using information, and obtaining the consent of the person whose data is gathered.
Network operators must not gather personal information unrelated to the services they provide; must not violate the provisions of laws, administrative regulations or agreements between the parties to gather or use personal information; and shall follow the provisions of laws, administrative regulations and agreements with users to process personal information they have stored.
Article 42 CCL: Network operators must not disclose, tamper with, or destroy personal information they gather; and, absent the consent of the person whose information was collected, must not provide personal information to others. Except, however, where it has been processed so that the specific individual is unidentifiable and cannot be recovered.
Network operators shall adopt technological measures and other necessary measures to ensure the security of personal information they gather, and prevent personal information from leaking, being destroyed or lost. When the leak, destruction or loss of personal information occur, or might occur, remedial measures shall be immediately taken, and provisions followed to promptly inform users and to make report to the competent departments in accordance with regulations.
Article 43 CCL: Where individuals discover that network operators have violated the provisions of laws, administrative regulations or agreements between the parties to gather or use their personal information, they have the right to request the network operators delete their personal information; where discovering that personal information gathered or stored by network operators has errors, they have the right to request the network operators make corrections. Network operators shall employ measures for deletions and corrections.
Article 44 CCL: Individuals or organizations must not steal or use other illegal methods to acquire personal information, and must not unlawfully sell or unlawfully provide others with personal information.
Article 45 CCL: Departments lawfully having network security supervision and management duties, and their staffs, must keep personal information, private information and commercial secrets, which they learn of in performing their duties, strictly confidential, and must not leak, sell, or unlawfully provide it to others.

- Consent of the person whose data is gathered (Art. 41, subsection 1)
- Disclosure of rules for the collection of data (Art. 41, subsection 1)
- Reasons for collecting data (Art. 41, subsection 2)
- Confidentiality obligations of network operators (Art. 42, subsection 1)
- An obligation on internet providers to provide notice to individuals, as well as reports to competent government departments, in cases where personal data have been leaked (Art. 42, subsection 2).

Although the NPC has enacted the Cyber Security Law, a number of deficiencies clearly still exist:

- This act does not represent an independent piece of legislation for the protection of personal data. Only six articles refer to the protection of personal data. Compared to the EU Data Protection Regulation, this regulation is too restricted in its scope.
- This law specifically regulates Internet operators. The rules on the collection, processing and use of personal data hardly refer to public bodies, or, respectively, the state administrative bureaucracy.
- In addition, the text does not contain any provisions stipulating competent independent supervisory authorities or a Data Protection Board headed by a supervisor comparable to the provisions contained in the EU Data Protection Regulation of 2016.

IV. The Outlook for Future Legislation on Data Protection in China

China is one of the most important economic powers in the world and has one fifth of the world's population. If the NPC continues in the long term not to pass stand-alone legislation to protect personal data, the personal rights of its citizens will not be protected, and the lack of such legislation will impede the international transfer of data to and from China.

The NPC should take the legislative experience of the EU and Germany in the field of data protection as a model, and at the same time take into consideration domestic needs and trends in development in the area of information technology. If it were to do this, it would create clear detailed principles, regulations and technical terms for the protection of personal data. In addition, China should establish a supervisory authority and data protection commissioners as the EU and Germany have done.

The protection of personal data during the period of rapid technological change today is an area of much debate. Technological developments such as Cloud Computing have frequently led to privacy issues in our society. It is obvious that such technological incursions can put personal privacy and the right to informational self-determination at risk. The traditional legal instruments for data protection may no longer be expedient or effective in the age of *Made in China 2025*.[34]

Both technology and the terms describing it are changing rapidly. This may be reflected, for example, in the shift in usage from *Big Data* to *Smart Data*.[35] That is why modern data protection law must be given a strong technological foundation. This is intended to provide the members of the legislature with a better understanding of the relationship between technology and data protection. In the normal scheme of things, data protection law always follows technology. The legislature reacts to a new technology with appropriate legal provisions, which are tailored specifically to the technology.[36] In order for the law to avoid being in a position where it has to react to the consequences of the introduction of new technology, new models of data protection such as "privacy by design" and "privacy by default" should be introduced. These models allow data protection requirements to be taken into account as early as the design stage of new technology.[37] "Privacy by design" is addressed to the manufacturers of products involving technology so that they can determine in advance which personal data their products will be able to collect.[38] The idea behind "Privacy by Default" is that data protection should be integrated as a system-based standard in order to limit to a required minimum the collection, processing and transmission of personal data from the very beginning of data processing.[39] It goes without saying that the other key principles of data protection law will be retained, including the principle of pro-

---

34 They will also no longer be appropriate in the future mobile world, cf. Hansen, "Zukunft von Datenschutz und Privatsphäre in einer mobilen Welt", *Datenschutz und Datensicherheit* 2015, p. 439.

35 Bretthauer, "Compliance-by-Design-Anforderungen bei Smart Data – Rahmenbedingungen am Beispiel der Datennutzung im Energiesektor", *Zeitschrift für Datenschutz* 2016, p. 268.

36 Schaar & Onstein, "Datenschutzrecht in der vernetzten Welt des 21. Jahrhunderts", *Bonner Rechtsjournal* 2011, p. 131.

37 Hilgendorf, "Automatisiertes Fahren und Recht", *53. Deutscher Verkehrsgerichtstag* 2015, p. 66.

38 Schaar & Onstein, *op.cit.* 2011 (Fn. 36), p. 131.

39 Schulz, "Privacy by Design", *Computer und Recht* 2012, 204.

hibition, the purpose-assigned principle, the principle of necessity, and the principle of transparency.[40]

In addition, modern data protection law should exist and be applicable at an international level. EU data protection law is a good example of this. Unfortunately, it would not be possible to enact similar legislation in the USA or China because of the particularities of the respective legal systems. That is why it is very important to compare data protection regimes, particularly the legislation of the EU and the USA.[41] China, when legislating in this area, should take advantage of all the experience gained so far in the EU and the USA. Finally, a uniform and convincing international data protection concept needs to be devised in order to establish an international minimum data protection standard.

F. The Example of Robotics and Law

Industrial robots will play an important role in *Made in China 2025*. The reason for this is the fact that traditional manufacturing industries in China no longer have a competitive advantage in terms of the cost of labor compared to Vietnam and India. In respect to the quality of goods produced, or the efficiency of production, China has no competitive advantage over Germany, the USA, or Japan. That is why Chinese industry is motivated to participate in the *Made in China* project. China hopes to further develop the use of robotics technology in order to increase the quality of its products and the efficiency of its production. In 2014 China became the largest consumer market of industrial robots in the world.[42] Nevertheless, the density of use of industrial robots in China is much lower than the world average. For foreign robot manufacturers, China will be an intriguing market in the future due to the enormous demand and government subsidies on the one hand. On the other hand, foreign robotics manufacturers face risks in the Chinese market because of the unclear and imperfect legal rules for the protection of intellectual property. In addition, the issue of

40 Bretthauer, *op.cit.* 2016 (Fn. 35), 271 *et seq.*

41 See also Orthwein & Rücker, "Kann Europa von Kalifornien Datenschutz lernen? ", *Datenschutz und Datensicherheit* 2014, p. 618; For details on the data protection conflict between the EU and the US, see Baumann, *Datenschutzkonflikte zwischen der EU und den USA*, 2016.

42 http://doc.xueqiu.com/14da51b1a6914103fe937026.pdf (in Chinese), 2015, p. 5.

tort liability for industrial robots has not been sufficiently clarified in current Chinese law.

In contrast, the criminal law in China is very well developed in comparison to foreign criminal law, including German criminal law. To illustrate this, I would like to solve a German case involving robots using the Chinese criminal law.

In June 2015, a technician was killed by a robot at the VW plant in Kassel.[43] The robot grabbed the worker, who was part of the team of people installing the robotics system, and hurled him against a metal plate. The preliminary investigation indicated that human error might possibly be responsible for the tragic accident. It was reported that the robot had been programmed so that it could perform various manufacturing tasks. The system was set so that it picked up auto parts and processed them further. On 7 October 2015 the Chinese newspaper *China Education Daily* reported on the accident using the headline "Robot Kills Man: Nightmare is coming".

Solving this case in Chinese criminal law involves the application of CLPRC § 134:[44]

Solution 1: If the victim was killed because the robot was incorrectly programmed by the staff responsible for that task, CLPRC § 134 subsection 1 may be applied. Under this provision, if employees violate safety regulations or rules, and, as a result, cause an accident involving severe injury or death, or other serious consequences, the employees shall be sentenced to fixed-term sentences of not more than three years. If the circumstances are particularly reprehensible, fixed-term sentences of 3 – 7 years may be imposed;

---

43 See http://www.computerwoche.de/a/volkswagen-roboter-toetet-mensch,3211777, 2015.

44 Article 134 CLPRC: Where anyone violates the provisions concerning the safety management in production or operations and thus causes any serious casualty or any other serious consequences, he shall be sentenced to fix-term imprisonment of not more than three years or detention. If the circumstances are extremely severe, he shall be sentenced to fix-term imprisonment of not less than 3 years but not more than 7 years.
Where anyone forces any other person to conduct risky operations by violating the relevant provisions so that any serious casualty or any other serious consequence is caused, he shall be sentenced to fix-term imprisonment of not more than five years or detention. If the circumstances are extremely severe, he shall be sentenced to fix-term imprisonment of five years or more.

Solution 2: If company management did not give the victim sufficient training to work safely with the robot, or forced the victim to work with the robot, then CLPRC § 134 subsection 2 applies. If company management force employees to work in an unsafe manner contrary to safety regulations, and as a result cause an accident involving severe injury or death, or other serious consequences, the respective managers shall be sentenced to fixed-term sentences of not more than five years. If the circumstances are particularly reprehensible, a fixed-term sentence of 5 years shall be imposed.

Solution 3: If the victim was killed by the robot because of human error, it is an accident.

G. Summary and Outlook

All in all, technology, as shown using the examples of hackers, cloud computing, an robotics in China in the context of *Made in China 2025*, raises a wide variety of legal issues that will occupy Chinese lawyers and politicians intensely in the coming years and decades. According to the five-year plan issued by the State Council, China has made significant achievements in recent years, but is still lagging behind in its innovation capabilities and information infrastructure. Additional funding will speed up the technological evolution of Chinese society. A number of specific areas of the law urgently require new legislation: the main focus of criminal law provisions against hacker crime needs to be shifted away from protecting computer systems towards ensuring better data security; stand-alone legislation to protect personal data should be passed by the NPC; the legal rules for the protection of intellectual property should be strengthened. In the context of the future widespread use of industrial robots, the NPC should now assume a forward-looking approach, and start considering how the law should deal with tort liability for injuries caused by industrial robots.

Technology helps people to pursue objective rules and to raise the efficiency of human labor. The law encourages people to be kind, to promote justice and preserve the social order. On the one hand, technology is used to fulfil human needs, but on the other hand, it always leads to changes in the structure of society, which makes legal reform necessary. When the law responds to technological developments, it is often necessary for it to deal with the negative consequences of those developments. Technology is constantly evolving. Correspondingly, the law must also constantly evolve.

China is now developing into a state under the rule of law. Reducing the negative impact of technological development and providing legal protection from the risks associated with that development are challenges facing all states under the rule of law.[45] The NPC will therefore need to integrate the changes both in technology and law. It has to adapt the law to take account of technological developments and it must ensure proper legal regulation of new technologies in order to ensure their safety and rapid introduction.

45 Cf. Haack, "Grundrechtsschutz und Risikoprävention – Eine verfassungsrechtliche Skizze", Jaeckel & Janssen, *Risikodogmatik im Umwelt- und Technikrecht*, 2012, p. 21 et seq.

New Legislation and the Reform of the Rules on Robots in Korea

*Minkyu Kim**,
Hanyang University, Seoul

A. The Current Situation in Korea

Some advanced industrial nations are currently suffering problems resulting from their low birth-rates and ageing populations. In this context, many Korean scientists and policy makers are suggesting that intelligent robots could be an effective solution to these problems.[1] Intelligent robots could be deployed to combat manpower shortages. Furthermore, social costs could be reduced by the popularisation of personal medical robots or household robots. Based on these prospects, the Korean government has adopted various laws and policies in respect of the robot industry. For example, in 2010 the Korean parliament adopted the Intelligent Robots Development and Distribution Promotion Act in order to provide state support for the development of the robot industry and to ensure the safety of robot products. Furthermore, a "basic plan for an intelligent robot" was set up and implemented under Article 5 of the Act. The first plan was implemented in the period 2009-2013.[2] 2014 saw the launch of the second plan, which aims mainly at acquiring essential technologies for robots and establishing a lead internationally in global robot markets.[3] It would therefore be useful to look at Korea's statutory approach as a way forward for legislation in the field of robot technology.

---

* Minkyu Kim studied law at Hanyang University in Seoul. He is currently working at the University of Würzburg; his research focusses on technology law and robotics and law.

1 The Korean Trade, Industry and Energy Ministry (ed.), *Research on the Actual Condition of the Robot Industry*, July 2011, p. 8.

2 The Korean Ministry of Education, Science and Technology (ed.), *Draft of first plan for an intelligent robot*, 2009, p. 25.

3 Park, in *TTA Journal* 2015, pp. 22-25.

In 2015, the Korean Institute of Criminology conducted a survey in order to gain information on the views of citizens about the new technology and the possibilities of introducing robots.[4] This survey asked how people viewed intelligent robots, what problems could derive from the introduction of robots, and to what extent robots should be employed in the field of politics and law.[5]

More than 80% of those surveyed felt that robots could be useful and helpful.[6] In contrast to this, they had little confidence in their safety: only 46.7% of those surveyed trusted robot systems.[7]

Regarding the intellectual capacities of robots, 86% of the general public and 73% of the experts surveyed said that intelligent robots can develop intellectually like children.[8] On the other hand, only 25% of the public and 17% of experts felt that robots' intellectual capacity could be deemed to be equal to that of an adult human being.[9]

With regard to the question as to who should be liable if harm is caused to a person's life, body or property by the operation of an intelligent robot, the majority (64%) of those surveyed said that the manufacturer of the robot should take responsibility. In particular, 77% of those surveyed said that the manufacturer should be liable even if the link between the outcome and the conduct of the robot could *not* be proven.[10] On the other hand, only 4.6% of the public and 2.4% of experts felt that the robot itself should bear responsibility for the outcome.[11] In contrast, 36% of those surveyed felt that users of robots should be liable if the cause of the harm and the objective assignment of this harm to the actions of the users were clearly proven and the outcome was clearly not caused by the manufacturer.[12] 86% of citizens and 90% of experts rejected the possibility of recognising basic rights for robots, and the majority said that irrespective of in-

---

4 More details in Yun, research into public views about the intelligent robot, cf. *Draft for advanced criminal law relating to the introduction of the new technologies*, pp. 193-295.
5 *Ibid.* (footnote 4), p. 197.
6 *Ibid.* (footnote 4), p. 289.
7 *Ibid.* (footnote 4), pp. 219-224.
8 *Ibid.* (footnote 4), pp. 225-231.
9 *Ibid.* (footnote 4), p. 231.
10 *Ibid.* (footnote 4), p. 231.
11 *Ibid.* (footnote 4), pp. 231-232.
12 *Ibid.* (footnote 4), pp. 234-239.

telligence, decision-making abilities and accountability, a strict distinction must be made between intelligent robots and human beings.[13]

Furthermore, many of those surveyed were concerned that powerful AI could deprive them of their jobs.[14] Unemployment is already increasing rapidly as a result of automated systems. Those surveyed were also concerned about the safety of robots and the possibility that their interests could be affected by crimes like the hacking of robots.[15] On the other hand, the majority of those surveyed said that intelligent robots could be used in public spaces to prevent crime, e.g. surveillance robots, patrol robots or guard robots could become a reality in the near future.[16]

This research suggests that the public, although it may be living in a different cultural context, generally has a positive view of robot technology. Nevertheless, the results show that the public attaches great value to human dignity, i.e. people regard robots merely as helpful and useful appliances which should be used in place of people to undertake difficult tasks. In addition, they are afraid of the dangers which could be caused by the introduction or application of new robot technologies. This attitude to robots is in line with what many scientists have already said. Lawmakers will need to pay careful attention to these views among the general public. There is therefore a need for new legislation or at least for the reform of existing rules.

---

13 *Ibid.* (footnote 4), pp. 240-243.

14 For example, in 2016 a famous Korean Go player lost against AI called AlphaGo (from Google Deepmind). Many Koreans were shocked by this and were scared by AI. Cf. Moyer, Christopher, "How Google's AlphaGo Beat a Go World Champion" in: *The Atlantic online*, 28 March 2016, available at: http://www.theatlantic.com/technology/archive/2016/03/the-invisible-opponent/475611/ (10 September 2016).

15 Yun, *op.cit.* 2015 (footnote 4), p. 292.

16 *Ibid.* (footnote 4), p. 293.

B. Legal Challenges posed by the Control of New Technology[17]

I. Legislation and an Ethics Charter for the Development of the Robot Industry

A Korean special committee presented the first draft of a *Charter on Intelligent Robot Ethics* in 2007. This charter is regarded by the Korean government as a benchmark for future legislation.[18] Furthermore, in 2010 the Korean legislature adopted the *Intelligent Robots Development and Distribution Promotion Act*, which entered into force on 1 January 2011. This act forms the basis for the development of an intelligent robots industry.

Although, in view of some contentious issues, the Korean legal system still needs to be modified and improved in the field of robot technology, it is remarkable that robot technology and the national robot industry were addressed in one go by the Act. Further to this, legislation relating to robot law was then revised to speed up the introduction of the new technology. In terms of fostering technological development, Korea's approach will probably prove itself to be very effective.

1. The Charter on Intelligent Robot Ethics

The first draft of the Korean Charter on Intelligent Robot Ethics was introduced by the Korean Minister for Industry and Energy to draw attention to the interrelationships between humans and intelligent robots and to make sure that the robot industry developed in the right direction.[19] The aims of the Charter are to establish an appropriate approach to intelligent robots and to give the general public easier access to the issue. Ultimately, it is about peaceful coexistence between people and intelligent robots. The Charter consists of the following chapters:

> *Article 1* (Aim): The aim of the Robot Ethical Charter is to confirm human ethics for the co-existence and co-prosperity of human beings and robots.

---

17 Unless stated otherwise, subsequent legal references are to Korean law.

18 Shin, "A practical and moral philosophical view of intelligent service robots" in *Korean Academic Society of Ethics*, 2010, p. 9 *et seq*.

19 Ko, "Principle of robot ethics: robot-related ontology" in *PAN-Korean Philosophy*, 2014, p. 401 *et seq*.

> *Article 2* (Common principles of human beings and robots): Human beings and robots must bear the dignity, information, and engineering ethics of life.
> *Article 3* (Human ethics): The production, use and handling of robots is only allowed within the limits of the law.
> *Article 4* (Robot ethics): Robots should obey human beings as a friend, helper and partner, and should not injure human beings.
> *Article 5* (Manufacturer ethics): The manufacturers of robots do so in order to defend the dignity of human beings, and are also responsible for robot recycling, and information protection duties.
> *Article 6* (User ethics): Robot users must respect robots as friends of human beings, and forbid any illegal reassembly of robots, as well as illegal robot misappropriation.
> *Article 7* (Promise of execution): The government and local authority organisations, in order to embody the spirit of the charter, must enforce the effective management of robot ethics throughout the manufacturing process and operation of robots.

The existence of intelligent robots can have a significant impact both on human lifestyle and on society. It can generate various risks, such as security problems with the robot and/or conflict between people and robots or between various robots.[20] The Charter on Robot Ethics was adopted in order to highlight the risks and to overcome the problems relating to robot technology and its related public concerns.[21] Even though the charter is not a piece of legislation and is therefore not binding, it is still important because governments would like to persuade their citizens that the related technology will be trustworthy due to the general rules, and that intelligent robots can be subject to ethical control. If the public trusts intelligent robots, the introduction and distribution of intelligent robot technology can be carried out more quickly.[22]

No fully autonomous ability to think has so far been developed in intelligent robots or in AI,[23] so that it is difficult to stipulate the specific content of legislation or to regulate specific aspects. Nevertheless, it is neces-

20 Shin *op.cit.* 2010 (footnote 18), p. 14 *et seq.*

21 Korea Institute for Robot Industry Advancement (ed.), journal Vol.7, of 15 April 2011, p. 8.

22 Shin *op.cit.* 2010 (footnote 18), p. 16 *et seq.*

23 Professor Oh, who developed a Korean intelligent robot called "Hubo", doubts whether an intelligent robot can have human-like capabilities or sensibilities. As a

sary to consider and avoid possible dangers associated with intelligent robots before the risks become a reality.

2. *Intelligent Robots Development and Distribution Promotion Act*

Article 1 of the Act states: "*The purpose of this Act is to contribute to enhancing the quality of life of citizens and the national economy by establishing and promoting a policy on the sustainable development of the intelligent robot industry to facilitate the development and distribution of intelligent robots and lay down the foundation therefor.*" According to Article 2 (1), an intelligent robot is "*a mechanical device that perceives the external environment for itself, discerns circumstances, and moves voluntarily*". In other words, the Act clearly regards intelligent robots as products. Article 2 (2) describes the definition and goal of the robot ethics charter.[24] The robot ethics charter contains principles on the development, manufacture and use of robots with a view to avoiding harm and preventing the violation of social order by robots. Article 2 (4) defines "robotland" as an area in which establishments and institutions are founded in line with Article 30 in order to accelerate the development and distribution of intelligent robots.[25] Government and local authorities are to draw up a

consequence, there is no need at present to reflect on legislation governing intelligent robots. Professor Ko has also said that the intelligent robot is merely an electronic appliance, meaning that it would be better to focus the Charter on the ethics of researchers, manufacturers or users of the intelligent robot. Even though intelligent robots can move independently, they cannot take autonomous decisions at present without human intervention. Robot ethics should therefore concentrate on the safety aspects of intelligent robots. From Kim, Un-hyon, "The robot: a tool or a partner?", *Hankook Media Network*, 29 December 2008, available at: http://daily.hankooki.com/lpage/coverstory/200812/wk20081229170638105450.htm (22 April 2014).

24 The title "Charter on Intelligent Robot Ethics" means a code of conduct established for persons involved in the development, manufacture, and use of intelligent robots in order to prevent various kinds of harmful or adverse effects, such as destruction of social order, that may arise from the development of functions and intelligence of intelligent robots, and to ensure intelligent robots contribute to enhancing the quality of life of human beings.

25 A *robotland* is an area hosting a robot R&D centre, a robot museum, a theme park and accommodation. Article 30 of the Act states: "(1) An area for the creation of a robotland shall be designated by the Minister of Trade, Industry and Energy at a request of the Special Metropolitan City Mayor, a Metropolitan City Mayor, a

budget plan pursuant to Article 3. They will make the necessary political decisions in order to accelerate the development and distribution of intelligent robots. Futhermore, the development and distribution of intelligent robots shall be based on this statute unless a more specific provision is applicable.

In order to ensure the quality of intelligent robots and to accelerate their distribution, under Article 9 (1): "*The Minister of Trade, Industry and Energy may establish policies necessary to secure the quality of intelligent robots and facilitate the distribution and dissemination of such robots.*" Article 17 states: "*The Government shall prepare measures necessary for facilitating the development and distribution of intelligent robots to improve convenience in using intelligent robots so that socially disadvantaged people, such as the disabled, the elderly, and low-income earners, can enjoy opportunities for and benefits from using freely such robots.*"

Metropolitan Autonomous City Mayor, Do Governor, or a Special Self-Governing Province Governor (hereinafter referred to as "Mayor/Do Governor", as prescribed by Presidential Decree. (2) When the Minister of Trade, Industry and Energy intends to designate an area for the creation of a robotland pursuant to paragraph (1), he/she shall consult in advance with heads of relevant central administrative agencies. (3) The cancellation of designation as an area for the creation of a robotland or an alteration to such area shall be made in compliance with the procedure for designation as an area for the creation of a robotland. In such cases, the consultation under paragraph (2) may be waived for any minor modification specified by Presidential Decree to such area. (4) The Minister of Trade, Industry and Energy shall, when he/she designates an area, cancels such designation, or alters the area pursuant to paragraph (1) or (3), publicly give notice thereof." On the other hand, the competent Mayor/Do Governor or a public agency specified by Ordinance of the Ministry of Trade, Industry and Energy may according to Article 31 (1) create a robotland in an area designated pursuant to Article 30 (1). The State or any local government may according to Article 34 fully or partially subsidize necessary construction costs and operating expenses only for the public facilities specified by Presidential Decree in developing a robotland.

C. Related Legislation

I. Acts on Road Safety[26]

1. Motor Vehicle Management Act

In 2016, the *Motor Vehicle Management Act* was amended in order to respond to the development of autonomous vehicle technology.

The definition of "autonomous vehicle" is set out in Article 2 (1)-(3) of the Act: a car that can move itself independently without the intervention or control of a driver or passenger. Anyone wishing to operate an autonomous vehicle for research or test purposes should file an application for a provisional licence to use the device under the provisions that apply with regard to safety and security checks, the district within which the test run is to be conducted, and the duties owed by the driver; this application pursuant to Section 27 (1) Sentence 2 should be filed with the Minister for Transport.

Further to this, pursuant to Article 29c (3) of the Act, manufacturers and dealers of vehicles containing an event data recorder (EDR) must provide the following information: 1. the recorded content from the EDR, 2. the study results, if the content from the EDR has been analysed. This provision aims to ensure that causation can be more easily proven between malfunctions in vehicles and accidents.

2. Road Traffic Act

So far, the term "the driver of a motor vehicle" has been used in various provisions,[27] i.e. a driver or a supervisor should be located in the vehicle despite the provisional test drive permit pursuant to Article 27 (1) sentence

26 For more details, cf. Kim, "Driverless transport systems: Present situation and legal problems of autonomous vehicles in Korea" in Hilgendorf, Eric & Hötitzsch, Sven (eds.), *Das Recht vor den Herausforderungen der modernen Technik,* Baden-Baden 2015, p. 149 *et seq*.

27 E.g. the Road Traffic Act contains the following provisions: Article 5: "Obligations to observe signals and instructions for pedestrians and drivers of motor vehicles"; Article 12 (3): "The driver of a motor vehicle shall pay attention to the safety of children near schools"; Article 12 (3): "The driver of a motor vehicle shall pay attention to the safety of the aged or the disabled in a protected area", and Ar-

2 of the Act to manage the vehicle in line with the road traffic rules. This means that the current Road Traffic Act cannot be directly applied to autonomous vehicles. For this reason, a revision of the Road Traffic Act is being discussed in Korea that would permit the operation of autonomous cars in public spaces.

II. Aviation Act and Aviation Ordinance Regarding Autonomous Aircraft

1. Definitions of Unmanned Aircraft and Ultra-Light Flying Devices

Pursuant to Article 2-3 (5) of the Aviation Act, an unmanned aircraft is defined as a flying device which does not need a pilot, but can "fly itself" by remote control or using an autonomous system. Further to this, Article 2 (8) of the Aviation Act in conjunction with Article 14 of the Aviation Ordinance says that ultra-light flying devices do not need pilots or passengers on board. According to Article 14-6 (1) of the Aviation Act, general unmanned aircraft should weigh less than 150 kg. Apart from this, unmanned aircraft can be viewed, pursuant to Article 2 (8) of the Aviation Act, as ultra-light flying devices. Such devices may be used without registration or approval as long as the ultra-light flying devices weigh less than 12 kg and are less than 7 metres in length.[28]

2. Commercial Uses for Ultra-Light Flying Devices

Article 16c (2) sets out the commercial uses of ultra-light flying devices: 1. support for agriculture, e.g. spraying fertilizer or spreading of seeds, etc.; 2. photography and filming for surveying or studying land and water; 3. observation of forests or public gardens; 4. pilot training. Despite this,

---

ticle 156 (1) "Penal provision for a driver of a motor vehicle who violates a provision", etc.

28 That means that ultra light flying devices are categorised as a general unmanned aircraft, and these types of devices are known as drones. Cf. Oh, "A normal use of intelligent robot technology and a criminological response" in Kim, Jin-Hwan (ed.), *Draft for advanced criminal legislation due to the introduction of the new technologies*, vol. 6, Korean Institute of Criminology, Seoul 2015, p. 402.

commercial operations shall not be permitted if they endanger life, health or property or might jeopardise national defence and security.

3. Compliance Conditions for Unmanned Aircraft

Pursuant to Article 23 (1) of the Aviation Act, owners must register their aircraft with the Ministry of Land and Transport. Following registration, owners must mark the respective aircraft registration numbers on the surface of their aircraft. If an unmanned aircraft is to be operated in a prohibited area, its user must obtain a corresponding permit from the Ministry of Land and Transport pursuant to Article 23 (2) of the Aviation Act. Pursuant to Article 23 (3), pilots of unmanned aircraft and ultra-light flying devices require written confirmation of being able to operate their devices.

Furthermore, pursuant to Article 23d the pilots of aircraft must fulfil special obligations if personal information can be collected during the operation of their unmanned aircraft and sent to others. In addition, pilots must report any air accidents caused by their aircraft. In order to protect the life and property of others, pursuant to Article 23 (8), Article 55 (5) of the Aviation Act and Article 68 of the Aviation Ordinance, pilots of unmanned aircraft must observe the following strict conditions: 1. The dropping of goods which can endanger the life or property of third parties is prohibited. 2. Flights shall be prohibited if unmanned aircraft are to be flown into densely populated areas or crowds of people and this can endanger the lives or property of third parties. 3. In accordance with Article 38 (2) of the Act, flight is prohibited in controlled spaces. Flights may only be permitted if the pilot and the operator have first received permission to fly from the director of a national airport. Further to this, the following activities can be permitted: a. use of ultra-light flying devices for military purposes; b. if aircraft weigh less than 12 kg or are less than 7 metres in length, they must remain below a maximum height of 150 metres; 4. Flights are prohibited where the pilot cannot see the device with his own eyes.[29] 5. Unmanned aircraft must be operated in daylight, i.e. in the period from sunrise to sunset.

---

29 Pursuant to Article 68 (4) of the Aviation Ordinance, a flight must always take place within direct view of the pilot. Alternatively, the pilot can obtain a permit pursuant to Article 23 (4) of the Aviation Act and remotely control the aircraft on that basis.

In the interest of air safety, pilots of ultra-light flying devices must give way to other aircraft pursuant to Article 68 (3). In addition, pursuant to Article 140 (2) of the Ordinance, pilots must check that their vehicles are in order before operating them. Whoever allows a flying unmanned aircraft or an ultra-light flying device to crash or be destroyed, shall be punished pursuant to Article 157 (1) of the Act. In addition, pursuant to Article 158 of the Act, the criminal punishment will be increased if the crash causes death or injury to human beings.

III. Medical Devices Act and Medical Service Act

It is necessary to discuss whether an intelligent robot can be recognised as a medical device.[30] A state-of-the-art medical device must first be granted permission under the Medical Devices Act by the Director of the Korean Food and Drug Administration (KFDA).[31] Further to this, the committee responsible for testing new medical technology[32] must, pursuant to Korean medical law, assess the reliability and effectiveness of novel equipment in order to safeguard the public health system and support the development of new medical technology.[33]

---

30 Lee, "Intelligent Service Robots – proposals for improvements in the law and the system", report to the Korean Robot Industry Association, *TTA Journal*, 2005, p. 55 *et seq*.

31 Medical Devices Act Article 6 (Permission, etc. for Manufacturing Business); Article 15 (Permission for Importation Business, etc.); Article 26 (Prohibition of Activities in General): "(1) No-one shall repair, distribute, lease, provide, or use a medical device without obtaining a permit or filing a report in accordance with Article 6 (2) or 15 (2), nor manufacture, import, repair, store, or display a medical device with intent to distribute, lease, provide, or use it".

32 Medical Service Act Article 54 (Establishment, etc. of Committee for Evaluation of New Medical Technology) "(1) The Minister of Health and Welfare shall establish the Committee for Evaluation of New Medical Technology (hereinafter referred to as the "Committee") within the Ministry of Health and Welfare for a review on the matters concerning the evaluation of new medical technology. (2) The Committee shall be composed of 20 or less committee members including one chairperson[...]".

33 Medical Service Act Article 53 (Evaluation of New Medical Technology) "(1) In order to protect public health and promote the development of medical technology, the Minister of Health and Welfare shall conduct an evaluation of the safety, validity, etc. of new medical technology (hereinafter referred to as "evaluation of new medical technology") through a review by the Committee for Evaluation of New

Telemedicine is another possibility. As robotic technology and information technology develop, telemedicine is growing in significance. In order to maintain the public healthcare system and to control telemedicine, the Medical Service Act was revised in 2010. Pursuant to Article 34 (1)[34] of the Act, a medical person[35] may use telecommunications technology or imaging communications to obtain support from a doctor at a different location who possesses specific medical expertise or skills. A doctor or a medical facility must provide suitable facilities and equipment for conducting telemedicine. A doctor who uses telemedicine technology to diagnose or treat patients must assume responsibility pursuant to Article 34 (3)[36] in the same way as in normal "face-to-face" treatment. If medical treatment is undertaken by a remote doctor, dentist or oriental medical doctor, and if a causal relationship between the doctor's telemedicine and a treatment error cannot be proven, the "local doctor" who is supported by telemedicine shall bear responsibility for the outcome pursuant to Article 34 (4).[37]

Medical Technology under Article 54, as prescribed by Presidential Decree. (2) "New medical technology" under paragraph (1) means medical technology newly developed, of which the Minister of Health and Welfare concludes necessary to evaluate the safety and validity. (3) The Minister of Health and Welfare shall notify the results of the evaluation of new medical technology to the President of the Health Insurance Review and Assessment Service under Article 57 of the National Health Insurance Act. In such cases, the results of the evaluation of new medical technology may be disclosed to the general public, as prescribed by Ordinance of the Ministry of Health and Welfare. (4) Other necessary matters concerning the subject, procedure, etc. for the evaluation of new medical technology shall be prescribed by Ordinance of the Ministry of Health and Welfare".

34 "Medical persons (limited only to medical doctors, dentists or oriental medical doctors who engage in medical service) may, notwithstanding Article 33 (1), give remote medical treatment (hereinafter referred to as "remote medical treatment") to furnish medical knowledge or technology to a medical person in a remote area by using information communication technology, such as computers or visual communication systems."

35 In this case, the medical person only means a doctor, dentist or oriental medical doctor.

36 "A person who gives remote medical treatment (hereinafter referred to as "remote doctor") shall have the same responsibility as when he/she gives direct medical treatment to a patient."

37 "If a medical person, who has performed medical practice following a remote doctor's remote medical treatment, is a medical doctor, dentist or an oriental medical doctor (hereinafter referred to as "local doctor"), and if there is no obvious ground to believe that the remote doctor is negligent in performing his/her medical prac-

1. Data Protection Law

The Personal Information Protection Act[38] sets out general data protection rules for Korea. Pursuant to Article 15 (1) of the Act, a user of personal information may collect, process and use this information if he first obtains the agreement of the owner of the information. The collection, processing and use of data are also permitted if it is necessary for a statutory purpose or for the work of competent authorities. Nevertheless, pursuant to Article 16 (1), as little information as possible shall be collected, and the use or transmission of the data may be restricted pursuant to Article 18 of the Act. In particular, the use of image processing devices[39] indoors, in which the privacy of individuals could be substantially infringed, is prohibited pursuant to Article 25 (2). This prevents collection of personal information which can occur via sensors or optical equipment of intelligent robots or unmanned aircraft.

Furthermore, the Act on Promotion of Information and Communications Network Utilisation and Information Protection, etc.[40] is in force. It focuses on data protection in the context of collection, use and transfer of private data which takes place via telecommunications and the Internet. The collection or use of personal information without the agreement of affected persons is prohibited pursuant to Article 22 of the Act. Further to this, Internet service providers or telecommunications service providers must take appropriate data protection measures pursuant to Article 28. Pursuant to Article 45 (1) of the Act, providers must take precautions to safeguard their networks and telecommunications. Pursuant to Article 3, protective measures must comprise the following safeguards: 1. physical and technical measures to prevent unauthorised or illegal access or to respond to such access; 2. technical measures to prevent an illegal outflow or illegal falsification and deletion of data; 3. physical and technical measures to ensure the continuous use of the network or the telecommunica-

---

tice, the local doctor shall be responsible for a patient, notwithstanding paragraph (3)."

38 "Personal Information Protection Act."

39 Article 2 (7). "Image processing devices are equipment which is permanently installed in a certain space in order to photograph a person or an object or to transfer images via a wired or wireless network."

40 Act on Promotion of Information and Communications Network Utilisation and Information Protection, etc.

tions; 4. administrative measures relating to human resources, structure and costs.

2. Civil Law

Pursuant to Articles 750 *et seq.* of the Korean Civil Code, whoever deliberately or negligently violates other persons' legal interests must pay damages. Thus civil liability exists in addition to criminal responsibility. Damages can be payable for breach of contract pursuant to Articles 390 *et seq.* of the Civil Code if a party fails deliberately or negligently to provide the required performance and his contractual partner suffers damage as a result. In addition, Korea has a Product Liability Act. Under Article 3 (1) of the Product Liability Act, manufacturers of products are obliged to pay damages to injured parties for personal injury, death or property damage due to faulty products. Intelligent robots are also industrial products, so this Act applies to robots. This Act is particularly important for the protection of consumers of intelligent robots,[41] but the Act also places a burden on manufacturers of robots because the Act basically imposes strict liability (liability without fault).[42]

3. Criminal Law

Manufacturers or users may incur criminal liability under Article 266 of the Korean Criminal Code if death, injury, or property damage are caused

41 Article 1 of the Product Liability Act: "The purpose of this Act is to help protect the injured persons and contribute to the improved safety of the citizens' life and the sound development of the national economy by providing for the liability of manufacturers, etc. for damages caused by the defects of their products."; Article 3 "(1) A manufacturer shall compensate for damages to the life, body or property of a person caused by a defect of a product (excluding damages inflicted only to the relevant product). (2) With respect to any product the manufacturer of which cannot be identified, a person who supplies it in a form of sale or lease, etc. for profit shall compensate for the damage referred to in paragraph (1), if he/she fails to inform any injured person or the legal representative of the injured person of the identity of the manufacturer or the supplier within a reasonable period, although he/she knows or would have been able to know the identity of the manufacturer or the person who supplied the relevant product to himself/ herself.".

42 For further details, cf. Kim, JF 2011, 59, 72-73.

by negligence. Here, the question arises as to whether a causal relationship can be proven between the manufacture or use of robots and damage.[43] Unlike German criminal law, in Korean criminal law the concept of negligence for a "manager" or an "occupation" exists. An "occupation" is an on-going operation or service exercised in the context of public life.[44] Case law states that driving a motor vehicle is an example of such an "occupation". That means that drivers who cause accidents by negligence and thus cause death or injury to others can receive more severe sentences under Article 268 of the Criminal Code than would be the case for general negligence.[45] This increased penalty is due to the fact that compared with other people, managers should pay more attention to their work, since due to their positions in business they have more information and expertise than laymen.[46] In contrast to this, Article 268 of the Korean Criminal Code also regulates gross negligence and imposes higher penalties in such cases.

IV. Proposals for Existing Legislation

1. Proposals to Change Existing Rules

Whilst current statutes and guidelines still contain a lot of preconditions and restrictions, some laws have been revised so that they apply to intelligent robots. One good example of this are the laws and ordinances relating to road traffic, which have been changed so that autonomous vehicles can now be used on public roads. In changing the law in this area, the Korean legislature has tried to respond quickly to the needs of industry and to identify potential problems at an early stage. Unlike the Korean government, EU Member States are displaying great caution in amending their norms. Making laws more flexible is not the best way forward, since irrespective of increased flexibility, legislatures and lawyers need to respond very quickly to the new challenges posed by robot technology.

---

43 Oh, *Criminal Law General Part*, 2nd edition, Seoul 2009, p. 195.

44 Oh, *ibid.* (footnote 43), pp. 195-196.

45 Article 268 A person who causes the death or injury of another by occupational or gross negligence shall be punished by imprisonment for not more than five years or by a fine.

46 Oh, *op.cit.* 2009 (footnote 43), p. 196.

In particular, it is feasible to harmonise safety standards for the manufacture and development of robot products, which are prescribed in various different laws and directives. In particular, there are no general standards at present for the programming of intelligent robots; rather, existing standards tend to be related to technical safety or the appearance of robot products.[47] On the one hand, this gap could stimulate public distrust of the use of robots in public spaces. On the other hand, developers or programmers of robots could move in the wrong direction because statutory and ethical restrictions or prohibitions on research have yet to be established. Correspondingly, a benchmark for the development and use of robot products should be considered. In this context, the degree of contact between people and robots could represent a significant benchmark for the safety rules to be obeyed by robots. Depending on the degree of contact, the safety standards could be graded into three classes.[48] For example, in class one robots do not undertake physical work with people. They will cause little danger and should therefore be subject to the safety standard for normal household appliances.[49] If an intelligent robot., e.g. a medical care robot, is used for a specific person, it would pose an average level of danger (class two). In this case, the conditions of use and physical strength could be restricted; also programmes would have to be installed to respond to emergencies.[50] If a fully automated robot can impact in public spaces on unspecified numbers of people (class three), e.g. a universal humanoid or an autonomous vehicle or aircraft, etc., a permit should be required from the competent authority for the robot's operation and it should be subject to regular inspections.[51] Furthermore, robots' artificial intelligence or programs must be designed with caution to ensure that programs do not violate existing laws and morals.[52]

47 For example, the Korean Trade, Industry and Energy Ministry has developed and adopted various safety benchmarks, e.g. the safety instructions for service robots (KS B 6935, 6936), the safety standards for service robots (KS B 6940, 6960, 6963), test standards on the durability of service robots (KS B 6961-6963) etc., but these standards refer to physical and technical safety.

48 Lee, *op.cit.* 2005 (footnote 30), p. 56.

49 Lee, *ibid.* (footnote 30), p. 56.

50 Lee, *ibid.* (footnote 30), p. 56.

51 Lee, *ibid.* (footnote 30), p. 56.

52 For example, Microsoft discontinued its "Tay" artificial intelligence service because of sexist and racist conduct. In addition, Google has apologised for the misconduct of its smart algorithms. It is difficult to prove whether these problems

2. Debate on Road and Air Traffic Safety

As described above, the rules contained in the Road Traffic Act and Aviation Act have been revised so that Korean industry can continue to develop unmanned systems.[53] It is expected that self-driving machines will become common in the near future. However, rapid changes in related legislation are still needed, since current provisions require either for human drivers to be present in autonomous vehicles or for aircraft to be deployed within sight of their controllers.[54] In particular, autonomous systems can be used directly in public spaces, so that they may endanger the safety of many unspecified persons.[55] It is feasible to define these systems as driver/pilot and to supplement the rules on traffic safety accordingly.[56]

Under such circumstances, it is necessary to require manufacturers to install black box-like event data recorders into autonomous systems and to supply consumers with the information recorded. This could facilitate proving causal relationships in accident situations.[57] Further to this, it is necessary to regulate in detail the substance of the duty of care which must

---

were caused by the AI algorithms themselves or by users. Nevertheless, it is apparent that even a program or algorithms can violate the moral or legal principles of human society. This means that the developers of algorithms need to pay close attention. Cf. Reese, Hope, "Why Microsoft's 'Tay' AI bot went wrong" in *Techrepublic*, of 24 March 2016, available at http://www.techrepublic.com/article/why-microsofts-tay-ai-bot-went-wrong/ (4 June 2016); Horton, Helena, "Microsoft deletes 'teen girl' AI after it became a Hitler-loving sex robot within 24 hours" in *Daily Telegraph*, 24 March 2016, available at http://www.telegraph.co.uk/technology/2016/03/24/microsofts-teen-girl-ai-turns-into-a-hitler-loving-sex-robot-wit/ (4 June 2016).

53 Kim, "The current situation and prospects of intelligent robot technology" in Kim, Jin-Hwan (ed.), *Draft for advanced criminal legislation due to the introduction of new technologies*, vol. 6, Korean Institute of Criminology, Seoul 2015, p. 71 *et seq.*; Oh *op. cit.* 2015 (footnote 28), p. 362 *et seq.*

54 Oh *ibid.* (footnote 28), p. 371 *et seq.*

55 Oh *ibid.* (footnote 28), p. 373 *et seq.*

56 More and more Korean experts are saying that the self-driving vehicle should be regarded as a driver, in order to facilitate the resolution of the possible legal conflict resulting from the recognition of the car's own responsibility. Cf. Kwon, Yong-Ju, "Is artificial intelligence a legal driver?", in *Autotimes*, of 12 February 2016, at: http://autotimes.hankyung.com/apps/news?mode=sub_view&popup=0&nid=01&c1=01&c2=02&c3=&nkey=201602102358561 (30 August 2016).

57 Kim (footnote 26), p. 166 *et seq.*

be met by owners, operators or controllers in the deployment of the unmanned systems. Specifically, operators or owners should only be able to use their machines for legal and moral purposes. In addition, they should be required to have their intelligent machines inspected regularly so that malfunctions or programming errors can be identified early and repaired. Furthermore, it will be necessary to amend existing insurance legislation in respect of self-driving machines.[58] For example, autonomous vehicles, in theory, are unable to act negligently, so there can be no talk of gross negligence like drunken driving or driving without due care and attention. Despite the general safety of autopilot systems, accidents can happen at any time, and accidents have indeed happened during the deployment of autonomous vehicles.[59] However, accidents might not only be due to autopilot error, but also to environmental circumstances or to the negligence of other road users and air travellers, so it is necessary for the legislature and insurance companies to develop new forms of insurance to handle such difficult situations.[60] In this respect, it would be expedient for safety standards to be defined unambiguously, taking account of the special features of intelligent robots and self-driving cars under the law. For example, it is feasible to require the installation of anti-hacking software to prevent unauthorised access or control of autonomous vehicles. It is also necessary for users and operators of intelligent cars not to be allowed to alter the structure or software of their vehicles without permission from the competent authority. Furthermore, the installation of an event recorder like a "black box" could be required so that causal relationships in accident situations can be proven more easily.[61]

3. Proposals for Telemedicine Robots

With regard to medical robots, telemedicine will likely play a major role in the future, since it will be possible very soon to undertake many diag-

---

58 For greater detail cf. Hilgendorf, "Automasiertes Fahren und das Recht" in *53. Deutscher Verkehrsgerichtstag 2015*, p. 66.

59 Cf. Oh, Won-Seok, "Googlecar has an accident: Google takes responsibility", of 2 March 2016, at: http://www.bloter.net/archives/251067 (30 May 2016); Chae, Ban-Seok, "Tesla Autopilot, a fatal accident has taken place", of 3 July 2016 at: http://www.bloter.net/archives/259098 (17 August 2016).

60 Hilgendorf *op.cit.* 2015 (footnote 58), p. 66; Oh *op.cit.* 2015 (footnote 28), p. 374.

61 Kim, *op.cit* 2015 (footnote 26), pp. 165-166.

noses or medical treatments using intelligent medical robots. In view of this debate, the Korean government is currently drafting medical legislation in order to uphold the health-related safety of patients treated via telemedicine.[62] The draft medical legislation aims to permit doctors to undertake telemedical treatment so that patients who either are in remote locations or are elderly or disabled can obtain high-quality medical services from qualified doctors.[63] The draft bill states that the establishment of a medical facility which aims solely to provide telemedical treatment is prohibited. If someone founds or manages such a facility, he would incur criminal liability.[64] Rather, doctors wishing to provide telemedicine services must first meet their new patients and carry out the first examination themselves. Furthermore, doctors must examine their patients in person at regular intervals.[65] Further to this, the provision of telemedicine could be restricted to non-fatal illnesses.[66] These proposals aim to protect patient health and to safeguard fair competition between medical facilities. On the other hand, under the draft legislation, the liability of remote doctors could be excluded in cases of erroneous treatment if patients have not followed their doctors' instructions, the damage was caused by patient error, or negligence on the part of the doctor was not proven.[67] According to the government's press release, the introduction of telemedicine has been successful, and patient satisfaction averages 85%.[68]

The draft bill is remarkable, on the one hand, because it requires that the preconditions and obligations put on medical facilities and doctors be

62 Kang, "Telemedicine between doctor and patient: draft amendment to law on doctors passed by cabinet", in: Korea Pharmaceutical Association News, of 7 June 2016, at: http://www.kpanews.co.kr/article/show.asp?idx=174315 (30 June 2016).

63 Press release / Ministry of Health and Welfare, "Draft change to law on doctors for the introduction of telemedicine", 10 December 2013, pp. 3-4, available at: http://www.mohw.go.kr/m/noticeView.jsp?MENU_ID=0403&cont_seq=294526&page=146 (30 June 2016).

64 *Ibid.* (footnote 63), p. 1.

65 *Ibid.* (footnote 63), p. 2.

66 *Ibid.* (footnote 63), pp. 2-3.

67 Ministry of Health and Welfare, "Expansion of telemedicine to promote the health service", press release of 2 August 2016, p. 9; at: https://www.bioin.or.kr/fileDown.do?seq=33827&bid=division (15 September 2016).

68 83% of patients living in remote areas and 88% of soldiers treated via telemedicine were satisfied. Also, 77.1% of sailors who engage in high-seas fisheries said that the telemedicine system was useful and necessary for them. Cf. Press release / Ministry of Health and Welfare, *op.cit.* (footnote 67), pp. 9-10.

set out clearly in law. But beyond this positive assessment, there is also a need to discuss the extent to which the practice of telemedicine should be allowed. At present, the law says that doctors may use telemedical equipment only for non-dangerous treatment. This provision could need to be amended again to accommodate future surgery robots. In addition, the current draft does not impose any standards relating to the safety of the machines used. Many doctors already expect to be able to use teleoperative robots to undertake surgical procedures. In such cases, the question of how doctors can respond to emergencies in the course of their operations, e.g. due to power outages or programming malfunctions, is particularly important.

It is feasible for safety standards and appropriate countermeasures to be defined by the Act. First, related acts, such as the Act on Medical Instruments or the Act on Intelligent Robots, need to be amended to prevent dangers to patient health. For example, international safety standards for intelligent robots and medical apparatus need to be implemented into domestic law. In addition, it is also feasible to require a medically qualified person to be present in a supervisory role in operating theatres in order to cope with potential emergencies.

4. Debate on Software in Respect of the Product Liability Act

In Korea, a chattle can be regarded as a product within the meaning of the Product Liability Act,[69] because "manufactured or processed" things are covered. This approach is typical, even though the Product Liability Act was actually influenced by U.S. rules.[70] The prevailing view in Korean legal science is that software is not yet a chattel covered by product liability under the existing act.[71] Correspondingly, the Korean Supreme Court has ruled that computer programs and information stored on data carriers cannot be regarded as things.[72] This fails to protect consumers who have suf-

69 Pursuant to Article 2 (1) a product is a movable object which is manufactured or processed.

70 Cf. Choi, in *Collection of scientific essays of Anseong Industrial University* 24 (2002), p. 58 *et seq*; Joo, in *The Korean Journal of Civil Law* 25 (2004), p. 434 *et seq*; Kim, in *Sungkyunkwan University Law Review* 16/1 (2004), p. 35 *et seq.*

71 In this regard Kim, *Justice* 1992, 20, 27; Lee, in *Academic Studies on Civil Law* 10 (2002), pp. 77-78.

72 Korean Supreme Court 2002 Do 745, 12 April 2002.

fered from damage caused by programming malfunctions. The potential of faulty systems to cause hazards will increase in the future because novel software systems will be designed as technology rapidly develops and is placed on the market.[73] In particular, the issue of artificial intelligence needs to be dealt with. Up to now, it has been possible for programs to be used in combination with physical or technical appliances, but AI could be used independently from cyberspace rather than from particular physical devices. It is therefore necessary to amend the Product Liability Act to cover software which consists only of non-physical information. The Korean government has already identified the problem and attempted to amend the Act.[74] In particular, this amendment could play a major role in terms of intelligent robots and self-driving cars which are largely dependent on programmes and AI.[75]

D. Conclusion

It is entirely up to the legislature to decide whether to give special priority to the development of modern technology and the national economy or rather to safeguard society and human dignity against novel risks. Certainly, the legislature needs to tackle the challenges posed by new technologies. Current laws and norms are not directly applicable to potential conflicts involving intelligent robot technologies, since their provisions were basically established to protect human dignity and the interests of people and society; this means that they are only indirectly applicable to robots

---

73 Kim, JF 2011, 59, 69 *et seq.*

74 In 2003, servers of an Internet provider were affected by a virus, and this caused damage to various authorities and companies. The damaged parties claimed that not only the Internet provider, but also programmers at Microsoft were liable because of the weakness of the Windows OS. So that a programmer of faulty software or an Internet provider can be held liable, a draft amendment to the Product Liability Act was presented by a committee of the Justice Ministry in 2012, but the draft was never adopted. Cf. Kim, Hunnam, "Amendment to the Product Liability Act: negative response from software engineers", in *Money-Today Online*, 23 August 2012, available at: http://www.mt.co.kr/view/mtview.php?type=1&no=2012082209454838498&outlink=1 (30 March 2015).

75 Cf. Lee, Ga-Young, "The software of a self-driving vehicle should be regulated under the Product Liability Act", in *Business Journal for Small Businesses*, 9 May 2016, available at: http://www.junggi.co.kr/article/articleView.html?no=14826 (20 May 2016).

via interpretations of standards and laws. It may therefore be expedient for the law to be changed in significant aspects. There are two approaches, either drafting new legislation or amending existing rules in order to accommodate robotic technology. Both offer advantages and disadvantages. Furthermore, the legislature needs to satisfy and affirm the needs of society and the legal system. Lawyers and policy makers need to tackle the practical issues relating to robotic technology without delay.

The Current Status of Japanese Robotics Law: Focusing on Automated Vehicles

*Takayuki Matsuo**,
Lawyer at Mom-o, Matsuo & Namba, Tokyo

Abstract

This paper outlines salient aspects of current Japanese robotics law. It reviews the most important legal discussions on robotics in Japan for the convenience of foreign (especially German and European) readers who are interested in Japanese robotics law. The paper then focuses on automated vehicles and summarizes the many important issues raised in Japanese legal discussions on automated vehicles. In the final section, it briefly examines criminal and civil liability for accidents involving automated vehicles from the viewpoint of Japanese law using the concepts of over-attribution and under-attribution.

A. Introduction

In the last few years, research on robotics law has increased significantly both in Japan and all over the world. The author is conducting research on robotics law, particularly from a comparative law perspective.[1]

---

* Attorney-at-law (admitted in Japan and New York), PhD candidate, Peking University Law School. Takayuki Matsuo would like to thank Professor Eric Hilgendorf for his support regarding the author's research.

1 The author made a presentation on Japanese robot laws in July 2016 in Shanghai, China. "The Preliminary Analysis on the Laws of Robotics in Japan -Using Automated Vehicles as Examples-" at IEE ARSO 2016 (https://ras.papercept.net/conferences/conferences/ARSO16/program/. *All the web resources were accessed on January 18, 2017.*). See also Takayuki Matsuo, *A Speech on Laws of Robotics at an International Conference in China,* 13(9) The Lawyers, 76 (2016). Then, in November 2016, the author had an opportunity to give a lecture on the current status of the research of laws of robotics from the perspective of comparative laws at

This paper first gives a view of the current state of Japanese research in robotics law. It then provides an overview of Japanese research on automated vehicles. Finally, the author briefly explains his opinions on the law related to automated vehicles, which obviously are from a Japanese legal perspective.

B. Research on Robotics Law in General

The Japanese government is trying to develop the robotics industry into one of Japan's main industries. Towards that end, it published the *New Robot Strategy* in February 2015.[2] At the same time, this area of research has become popular among Japanese legal scholars and practitioners. The Study Group of Law of Robotics was established in 2016 at the Information Network Law Academy.[3]

Below is a general summary of research on robotics law in Japan as of January 2017.

I. Research by the Ministry of Internal Affairs and Communications

One of the most notable activities of the Japanese government in this field is the research conducted by the Institute for Information and Communications Policy of the Ministry of Internal Affairs and Communications. In particular, their work on the regulation of research and development in artificial intelligence (or AI) is quite important. In 2015, the Institute organized a series of conferences called the Study Group Concerning the Vision of the Future Society Brought by Accelerated Advancement of Intelligence in ICT. After publishing its midterm report in June 2015,[4] it was renamed as the AI Networking Review Conference. It published its

---

the Information Network Law Association in Japan. See http://windy.mind.meiji.ac.jp/InLaw2016/.

2 http://www.meti.go.jp/press/2014/01/20150123004/20150123004b.pdf.

3 http://www.in-law.jp/bn/2016/20160331-2.html.

4 Study Group concerning the Vision of the Future Society Brought by Accelerated Advancement of Intelligence in ICT "Report 2015" (Released on June 30, 2015) (English Abstract) (http://www.soumu.go.jp/main_sosiki/joho_tsusin/eng/pdf/151030_01.pdf) "Report 2015" (Japanese full version) (http://www.soumu.go.jp/main_content/000363712.pdf).

midterm report in April 2016[5] and its final report in June 2016.[6] One of the most significant items in the report was a proposal for eight guiding principles for the research and development of AI, namely the principles of transparency, user assistance, controllability, security, safety, privacy, ethics, and accountability. This was also discussed during the G7 ICT Ministers' Meeting in April 2016.[7] Thereafter, the Institute organized a further study group called the Study Group on the Promotion of AI Network Society in October 2016. In December 2016, the Institute published a call for public comments on its guidelines for research and development in artificial intelligence.[8] Public comment ended in January 2017 and it is expected that the Study Group will consider the results of the public consultation and prepare draft AI Research and Development Guidelines which will then be put forward for international discussion.

II. Summary of Foreign Research

We should admit that the present state of research in Japan is falling behind Europe and US. Therefore, Japanese scholars are interested in the results of foreign research in robotics law, especially German research.

One of the most notable Japanese efforts is the study group called "Robot and Criminal Law" held by Tetsuya ISHII of Chiba University. Professor ISHII and other members of the study group are translating and summarizing the articles published in the *Robotik und Recht* (Robotics and Law) series, edited by Professor Eric Hilgendorf.[9]

---

5 "Midterm Report" of Study Group of AI Networking (press release) (http://www.soumu.go.jp/menu_news/s-news/01iicp01_02000049.html).

6 "Report 2016" of Study Group of AI Networking (press release) (http://www.soumu.go.jp/menu_news/s-news/01iicp01_02000050.html).

7 Proposal for Discussion toward Formulation of AI Research and Development Guideline (http://www.soumu.go.jp/joho_kokusai/g7ict/english/main_content/ai.pdf).

8 Call for public comment on Discussion Points as to the Creation of the Draft for International Discussion on the Formation of the "AI Development Guideline" (https://search.e-gov.go.jp/servlet/Public?CLASSNAME=PCMMSTDETAIL&id=145208852&Mode=0).

9 Kosuke IMAI, "Summary and Comments", 31(2) *Chiba Journal of Law and Politics* 97 (2016) (reviewing Jan C. Joerden, "Strafrechtliche Perspektiven der Robotik" in Eric Hilgendorf & Jan-Philipp Günter (eds.), *Robotik und Gesetzgebung* (2013) p. 195), Masamitsu TOMIKAWA, "Summary and Comments", 31(2) *Chiba Jour-*

In addition, there are papers introducing German law,[10] American law,[11] French law,[12] and so on.[13]

---

nal of Law and Politics 132 (2016) (reviewing Eric Hilgendorf, "Recht und autonome Maschinen – ein Problemaufriss" in Eric Hilgendorf & Sven Hötitzsch (eds.), *Das Recht vor den Herausforderungen der modernen Technik* (2015) p. 11, and Yoshiaki ITO, "Summary and Comments", 31(2) *Chiba Journal of Law and Politics* 138 (2016) (reviewing Eric Hilgendorf, "Können Roboter schuldhaft handeln?" in Susanne Beck (ed.), *Jenseits von Mensch und Maschine* (2012) p. 119).

10 Georg Borges (tr. T. SUIZU), "Pepper and Law – Is it necessary for Law Reforms in Order to Tackle with New Information Technology?", 89 (7) *Journal of Law, Politics and Sociology* 57 (2016); Susanne Beck (tr. M. TOMIKAWA), "Robotics and Law: The Problem, Current Discussions, and First Clue for Solution", 50(2) *Journal of Comparative Laws* (hikakuho-zasshi) 95 (2016); Kyoko KANEOKA, "Automated Vehicle and Civil Liability in Germany", 1501 *Jurist*, 44 (2017); and Eric Hilgendorf (tr. M.TOMIKAWA) „Automatisiertes Fahren und Recht", 51; *Journal of Comparative Laws* (hikakuho-zasshi) (2017).

11 Susumu HIRANO, "eBay, Inc. v. Bidder's Edge, Inc", 28(7) *International Commercial Law* (kokusai-shoji-homu) 898 (2000), Susumu HIRANO, "Mracek v. Bryn Mawr Hospital", 44(6)) *International Commercial Law* (kokusai-shoji-homu), 956, Susumu HIRANO, "Payne v. ABB Flexible Automation, Inc.", 44(7) *International Commercial Law* (kokusai-shoji-homu), 1114 (2016), Susumu HIRANO, "Miller v. Rubbermaid, Inc.",44(9) *International Commercial Law* (kokusai-shoji-homu) 1430 (2016), Katsunori KAI, "Nanotechnology and Medical Robotics: Legal and Ethical Responsibility", 30 Waseda Bulletin of Comparative Law 1 (2012), Mieko ITO, Information on Legislation: FAA Draft Drone Rules for Commercial Use, 263(2) Foreign Legislation (gaikokuno-rippo) 6 (2015), Gen GOTO, "Automated Vehicle and Civil Liability in the U.S.", 1501 *Jurist*, 50 (2017), and Koji TOJIMA et. al., Comparison of Japanese and Chinese new Drone Regulations. 1084 NBL 42 (2016).

12 Larrieu Jacques, "Androïdes, exosquelettes, prothèses et corps humain: Une tentative de definition d'un statut des robots" (French law, 62 *Osaka University Law Review* 73 (2015) and Larrieu Jacques (tr.Y. SAITOU, "Android, Powered Suits, prosthesis and human body", 64(1*) Osaka University Law Journal* (handaihogaku) 301 (2014).

13 Yoshihiro Kusuda, "Book Review", 231 *Robot* 41 (2016) (reviewing Robert Bogue, "Robot ethics and law: Part two: law", 41(5) *Industrial Robot: An International Journal*, 398 (2014)) and Ayumi MITSUNARI, "Information on Legislation: Singapore Enforced a Law to Promote Safety Use of Drones", 264(1) *Foreign Legislation* (gaikokuno-rippo) 24 (2015).

III. Artificial Intelligence

Historically, research on artificial intelligence and the law already started in the 1980s. However, the focus of research at that time was from the viewpoint of programming computers to understand and mimic legal reasoning.[14]

Recently, in contrast, research on legal issues related to artificial intelligence has become popular. Some of the issues are the copyright of works "created" by AI,[15] civil liability,[16] FinTech and use of AI in the financial industry,[17] issues in criminal investigation and criminal procedural law,[18] etc.[19]

---

14 Akihiko NAKAMURA, "Application of Artificial Intelligence to Law" (1)(2), 2 and 3 *Kinki University Journal of Industry and Legal Information Institute* (Kinki Daigaku Sangyou to Houjouhou Kenkyujo Kiyo) (1988). Hajime YOSHINO, "Development of Legal Expert System – Legal Reasoning by AI", 97 *Private Law Information*, 6 (1994); Yoshiyuki MATSUMURA & Shigeru KAGAYAMA, *Legal Artificial Intelligence – Clarification of Legal Knowledge and Embodiment of Legal Reasoning*, (2000) and Seiichiro SAKURAI, "Artificial Intelligence and Legal Artificial Intelligence", *29 Yearly Report of Legal Science Institute,* Meiji Gakuin University (houritsu-kagaku-kenkyujo-nenpo) 1 (2013).

15 Kensaku FUKUI, "Artificial Intelligence and the Copyright 2.0: How the Copyright System Changes Due to Increase in Creation of Works by Robots", 652 *Copyright* 2 (2015) and Seiji SUGIMOTO, "Do AIs 'Dream'"? 648 *Copyright* 1 (2015).

16 Hirofumi ITO, "On the Civil Liability of Artificial Intelligence", 206 *Legal and Economic Journal of Aichi University School of Law* (Aichi-daigaku-hogakubu-houkei-ronshu), 67 (2016).

17 Hiroyuki SANBE and Takafumi OCHIAI, "Legal Issues regarding the Use of AI by Financial Institutions", 67(37) *Weekly Financial Affairs* (shukan-kinnyuu-zai-sei-jijo) 32 (2016).

18 Shunichi KAWABE and Yoshitaka SHIMADA, "A Study on the Method of Digital Forensics for the Deployment of IoT and Artificial Intelligence", 69(5) *Journal of Police Study* (University of Police) (Keisatsu-gaku-ronshu), 54 (2016).

19 Naohiko OGIWARA, "Tackling the Research and Development of AI by Ministry of Internal Affairs and Communications", 58(6) *Jihyo* 184 (2016); Takafumi OCHIAI *et. al.*, "Summary of Legal Effects of IoT, AI, Sharing Economy and Development of IT", 13 (8) *The Lawyers* 56 (2016); Hirofumi ITO, "On Legal Singularity", 26(1) *Com* 13 (2016), and Nobuhiko HADANO, "'SINGULARITY' and Law", 12 *Journal of Hiroshima Law School* (Hiroshima-hoka-daigakuin-ronshu), 39 (2016).

IV. International Public Law

Another important research field for recent robotics law is international public law, especially on the topic of autonomous lethal weapons. There are a relatively large number of articles by Seigo IWAMOTO[20] and Heigo SATO.[21]

V. Drones

One of the most important events affecting the legal regulation of drones in Japan was the 2015 Tokyo Drone Incident. It transpired that a Phantom 2 drone carrying traces of radiation was found at the Japanese Prime Minister's Official Residence in April 2015. After this incident, there were calls for the strict regulation of the use of drones. In December 2015, an act to amend the Civil Aeronautics Act was passed which strictly regulates the use of drones.

Subsequently, many articles on drones and the legal regulation of drones have been published.[22] Furthermore, the Public-Private Sector

---

20 Seigo IWAMOTO, "Memorandum on the Legality of UCAV from International Public Law", 45(3/4) *Sandai Hogaku*, 685 (2012); Seigo IWAMOTO, "A New Trend Towards Development of an International Legal Framework to Regulate Lethal Autonomous Robotics (LARs)", 47(3/4) *Sandai Hogaku* 330 (2014); Seigo IWAMOTO, "International Legal Assessment of Unmanned Weapons and the Trend towards Their Regulations", 42(2) *International Security* (kokusai-anzenhosho), 15 (2014); Seigo IWAMOTO, "How to Deal with Coming Robotic War, Start of the Discussion on Legal Regulation of LARs", 104 *Komei*, 46 (2014); Seigo IWAMOTO, "Drones and Legal Regulation", 35(7) *Journal of Defense Technology* (bouei-gijutu-journal) 16 (2015); Seigo IWAMOTO, Book Review, 44(1) *International Security* (kokusai-anzenhosho), 98 (2016) (reviewing Steven J. Barela (ed.), Legitimacy and Drones: Investigating the Legality, Morality and Efficacy of UCAVs, (2015)), and Seigo IWAMOTO, "Small Drones Laws in Japan: In Comparison to Other Countries", 161 *CISTEC Journal* 82 (2016).

21 Heigo SATO, "Lethal Autonomous Robotic Weapons", 62(2) *Journal of World Affairs* 93 (2014) and Heigo SATO, "What to Worry About the Lethal Autonomous Weapon Systems (LAWS)", 42(2) *International Security* (kokusai-anzenhosho) 1 (2014).

22 Symposium, *The Legal Response and Liability Issues on the Safety of Robots and Drones, Record of 14th Conference of Information Network Law Academy* 33 (2015), Hiroshi MIYASHITA, "Drones and Privacy", 1975 *Law Times* (tokinohorei), 54 (2015), Mayu TERADA, "New Laws Explanation: The Amendment of

Conference on Improving the Environment for UAVs (April 2016) produced the so-called *Roadmap for the Application and Technology Development of UAVs in Japan.*[23] This document provides a framework for future rulemaking and policymaking on drones by the Japanese government.

VI. Other Research

There are other works on robotics law but one important contribution would be the series of publications by Fumio SHIMPO in *Law Times* (*tokino-horei*), which basically covers all fields of robotics law.[24]

---

the Civil Aeronautics Law and the Introduction of the Regulations on Drones", 426 *Hogaku Kyoshitsu*, 47 (2016), Aviation Safety and Security Planning Division of Aviation Safety and Security Department of Ministry of Land, Infrastructure, Transport and Tourism, *On the Flying Rules of Unmanned Aerial Vehicles including Drones, 80 Small Airplanes and Safe Operation* (kogatakito-anzen-unko) 2 (2016). Daisuke NAKABAYASHI, "Laws on Amendment of Civil Aeronautics Law", 413 *Horei Kaisetsu Shiro Soran* 31 (2016), Daisuke NAKABAYASHI, "Law Explanation: On the Creation of Flying Rules of Unmanned Aerial Vehicles including Drones – Laws on Amendment of Civil Aeronautics Law", 52(8) *Construction Machines* (kensetsu Kikai) 1 (2016), Yoshiyuki ASO, "Safety Measures for Usage of Drones", 22(4) *Monthly Safety Forum* (gekkan-chian-forum) 2 (2016), Koji TOJIMA and Hiromi HAYASHI, "Business Viewpoint for Drone Regulation", 15(10) *Business Legal Affairs* (Business-houmu), 4 (2015), Yaichi YAMADA, "A Study on the Drone Business and Legal Regulation", 6 The *Journal of Business Management*, 9 (2016), Hiromi HAYASHI, "Q&A Corporate Legal Affairs : Regulation on Drone by Amended of Civil Aeronautics Law", 68(2) *Accounting* 275 (2016), Ryo TAKIZAWA, "Legal Information Basics, Drone Laws and the Points of Revisions", 8(12) *Business Law Journal* 68 (2015), Takashi NAKAZAKI, "Current Status of Drone Regulation", 1061 *NBL* 26 (2015), Soichi KUDO, "Security Regulation on Drones – On the Law to Ban Unmanned Aerial Vehicles", 21(10) *Safety Forum* (chian-forum) 27 (2015), Tamiko YABUKI, "Drones and Insurance Industry: Taking American Example as Guidance", 117 *Sonpo-soken Report* 1 (2016).

23 http://www.meti.go.jp/policy/mono_info_service/mono/robot/downloadfiles/uasroadmap_ENG.pptx.

24 Fumio SHIMPO, "What to Learn from Abroad Regarding the Policy on Laws on Robotics", 2001 *Law Times* (tokino-horei) 2 (2016), Fumio SHIMPO, "What is Robot", 2003 *Law Times* (tokino-horei) 2 (2016), Fumio SHIMPO, "Autonomy of AI", 2005 *Law Times* (tokino-horei) 2 (2016), and Fumio SHIMPO, *Problems to be Discussed regarding the Laws of Robotics from the Viewpoint of Legal Fields(*1)-(6), 2007-2017 (tokino-horei) 2 (2016-2017).

In addition, there are works on civil law,[25] criminal law,[26] consumer law,[27] telecommunication law,[28] IP law,[29] and other fields.[30]

Finally, there is one area of robotics law which has been under discussion for quite some time. That is the law on industrial robots, especially from the viewpoint of labor law. Regulations in health and safety have existed for decades to protect employees from potential risks posed by industrial robots.[31] However, as the focus of such research is mostly on the interpretation and practice of black letter legal rules, this article will not focus on this area.[32]

---

25 Susumu HIRANO, "Ensuring Safety of Robots – PL and Safety Standard", 61(8) *Study on Urban Problems* (toshi-mondai-kenkyu) 80 (2009), Ryota AKASAKA, "Study on 'PL of Information'", 14 *Information Network Law Review* 129 (2016), Ryota AKASAKA, "PL of Automated Robots" 13(2) *Information Network Law Review*, 103 (2014), and Ryota AKASAKA, "Torts Liability on Provision of Tele-existence", 12 *Information Network Law Review*, 1 (2013).

26 Akihiro ONAGI, "Times of Scientific Technology and the Form of Criminal Law: The Advocacy of Cyborg Criminal Law", 63 (6) *Hokkaido University Law Review* (hokudai-hogaku-ronshu), 2152 (2013).

27 Tomoko NAMBU, "Legal Knowledge on Robots – Consumer Product Safety Act and Product Liability Law", 458 *Consumer Information* (shohisha-joho) 8 (2015).

28 Hiromi HAYASHI, "Legal System on the Usage of Radio Waves", 1074 *NBL* 42 (2016).

29 Yumiko WASEDA, "Choreography of Robots", 638 *Copyright*, 37 (2014).

30 Fumio SHIMPO, "The Beginning of Robot Laws", 27 *Nextcom*, 22 (2016) Ikuo TAKAHASHI, "Legal Risk Legal Issues of the Next Generation of Robots", 13(1) *Research and Development Leader* (kenkyu-kaihatsu-leader), 54 (2016), Masahiro KOBAYASHI, "New Generation of Robots and the Society in the Future", 458 *Consumer Information* (shohisha-joho) 6 (2015), Masahiro KOBAYASHI, "Rule of Industry Law and the Form of Japan from the Viewpoint of Robots", 67(2) *Liberty & Justice* 8 (2016), and Tomoo SHIBANO and Hironori KOBAYASHI, "Caring Robots and Legal Regulation", 15(8) *Business Legal Affairs* (Business-houmu), 103 (2015).

31 See for example, *Japan Industrial Safety and Health Association, The Theory and Practice of the Safety Management of Industrial Robots* (2nd ed. 2016), and Junichiro SHIRASAKI, *100 Questions and Answers on Industrial Robots* (2016).

32 However, there are some who focus on new aspects of industrial robots and law. See Shinya OUCHI, *Laws and Work Life in AI Era* (2017) Tomoko NAMBU and Takuro HASHIMOTO, "Legal Problems on Robots and Effects to Laborers", 65(11) *Science of Labor* (rodo-no-kagaku) 19 (2010) and the author's speech on Japanese robot labor laws in the new era at the Japanese Robot Law Society Preparatory Conference in October 2015 (see https://wirelesswire.jp/2016/01/49117/).

C. *Discussions of Automated Vehicles and the Law*

I. Recent Discussions of Automated Vehicles in Japan

One of the most important fields of robotics law in Japan is in respect to automated vehicles. Widespread use of such vehicles would bring many advantages, such as a decrease in the number of car accidents, increased mobility for the elderly and the disabled, efficient and environmentally friendly driving, etc.[33] Partially automated cars have already begun to be sold in Japan.[34] Field demonstrations of automated cars are widely conducted by many Japan-based automobile companies. At the same time, many Japanese people and the government are worried by car accidents caused by partially automated cars.[35] Furthermore, some insurance companies have started to sell (non-compulsory) automobile insurance which covers accidents caused by automated vehicles.[36]

The Japanese government has the goal of developing fully automated vehicles. *Roadmap 2016 Government and Private Sector ITS Plan*,[37] which was published by the Japanese Prime Minister's Office, discusses the planned implementation of different levels of automated vehicles.[38]

---

33 Eric Hilgendorf, "Teilautonome Fahrzeuge: Verfassungsrechtliche Vorgaben und rechtspolitische Herausforderungen" in Eric Hilgendorf, Sven Hötitzsch & Lennart S. Lutz (eds.), *Rechtliche Aspekte automatisierter Fahrzeuge,* 16 (2014).

34 See, for example, the new Nissan Serena on sale August 2016, which had a partially automated function called "ProPILOT." (http://nissannews.com/en-US/nissan/usa/releases/nissan-s-new-serena-propilot-technology-makes-autonomous-drive-first-for-japanese-automakers).

35 For example, after the tragic accident by Tesla Motors' partially automated car, the Japanese Ministry of Land, Infrastructure, Transport and Tourism made an announcement that users of "automated vehicles" should be careful as there is no "fully" automated vehicle available on the market and the "automated" functions are only for assistance of the drivers. (http://www.mlit.go.jp/common/001137302.pdf) A similar point was raised by Hidehiro WATANABE, "Current Status of Tackling Automated Driving and the Issues of Insurance", August 2016 *Kyosaisoken Report* 16 (2016) (http://www.jkri.or.jp/PDF/2016/Rep146watabe.pdf).

36 See Yuusuke IKEDA, "Current Status of Automated Driving Technology", 1501 *Jurist*, 21 (2017).

37 http://www.kantei.go.jp/jp/singi/it2/kettei/pdf/20160520/2016_roadmap.pdf..

38 According to the Roadmap, Level 3 Autopilot was planned to be embodied in 2020. Level 4 automated vehicles are separated into two categories. Level 4 automated vehicles with remote control or in a limited space (such as a certain area of

Readers are also referred to the *Japan Revival Plan 2016*, approved by the cabinet in June 2016, which decided to develop fully automated vehicles.[39]

Given the high level of interest in this area, several Japanese ministries have organized series of research projects on automated vehicles. In 2013, the Ministry of Land, Infrastructure, Transport and Tourism organized a study group on auto-pilot systems, and in October 2013, the *Mid-term Report for the Embodiment of the Auto-pilot Systems* was published.[40] Thereafter, in 2015 the Ministry organized the Study Group on the Automated Mobility Business and in February 2016 it published the Report entitled *The Future Policy*.[41] In 2015, The National Police Agency also organized a Committee on the Study of the Systematic Issues of Automated Mobility, and in March 2016 the Committee published the *Report on the Study of the Systematic Issues of Automated Mobility*.[42] Based on the report in May 2016, the *Guidelines for Field Experiments of Automated Mobility Systems* were published.[43]

Research on automated vehicles and the law has also become popular. In November 2016, the Study Group of the Liability of Compensation for Damages regarding the Automated Driving commenced work. It is sponsored by the Ministry of Land, Infrastructure, Transport and Tourism.[44] Furthermore, in June 2016, the General Insurance Association of Japan published the report entitled *On the Legal Issues of the Automated Driving*.[45]

---

highways) are planned to be embodied in 2020 and fully automated vehicles without any limitation are planned to be embodied in 2025.

39 http://www.kantei.go.jp/jp/singi/keizaisaisei/pdf/2016_zentaihombun.pdf and http://www.meti.go.jp/committee/sankoushin/shin_sangyoukouzou/009_haifu.html.

40 http://www.mlit.go.jp/road/ir/ir-council/autopilot/pdf/torimatome/honbun.pdf (interim summary).

41 http://www.mlit.go.jp/common/001124331.pdf.

42 https://www.npa.go.jp/koutsuu/kikaku/jidosoko/kentoiinkai/report/honbun.pdf.

43 www.npa.go.jp/koutsuu/kikaku/gaideline.pdf.

44 http://www.mlit.go.jp/report/press/jidosha02_hh_000268.html.

45 http://www.sonpo.or.jp/news/file/jidou_houkoku.pdf.

Given that there are many publications on legal issues related to the use of automated vehicles,[46] the next sections will summarize the arguments respectively in the civil law and the criminal law.[47]

Of particular interest in respect of the allocation of liability for automated vehicles are discussions about levels of automation. One of the most widely used classification systems was defined by the American Federal National Highway Traffic Safety Administration (NHTSA), which used levels 0 to 4.[48] Another is SAE International's definition, which used levels 0 to 5.[49] Although discussions referred to in this article frequently use levels of automation, it should be kept in mind that levels do not necessarily allow conclusions to be drawn on legal liability and can sometimes be misleading.[50]

II. Discussions on Automated Vehicles and the Civil Law

Looking at Japanese tort law, § 709 Japanese Civil Code states: "a person who has intentionally or negligently infringed any right of others, or legally protected interest of others, shall be liable to compensate any damages resulting in consequence". When a driver is negligent and causes an accident, he or she is liable for the damage resulting from such an accident. In

---

46 See Ikuo TAKAHASHI and Mayu ARIMOTO, "Legal Issues of Automobile System – Focusing on Automated Driving", 14 *Information Network Law Review*, 112 (2016), Koji TOJIMA, "The Current Status and Issues of the Legal System for the Embodiment of the Automated Vehicle (Automated Driving)" (1)(2), 1073 *NBL* 28 and 1074 *NBL* 49 (2016), Koji NAKAYAMA, "Legal Issues and Liabilities of Automated Driving", 11(2) *Hitachi Soken*, 28 (2016), Koji NAKAYAMA, "Legal Issues regarding the Automated Driving", 69 (12) *Journal of Society of Automotive Engineers of Japan*, 39 (2015), Takeyoshi IMAI, Legal Issues regarding the Automated Driving, 40(2) IATSS review 40(2), 132 (2015).

47 One of the important articles which cannot be categorized into civil and criminal laws would be Susumu HIRANO, "Robot Law and the Derivative of Trolley Problem of Automated Driving", 1083 *NBL* 29 (2016), which discussed a similar problem to the algorithms of death (Algorithmen des Todes) presented by Eric Hilgendorf, *loc. cit.* (f. 30).

48 National Highway Traffic Safety Administration, Preliminary Statement of Policy Concerning Automated Vehicles (http://www.nhtsa.gov/staticfiles/rulemaking/pdf/Automated_Vehicles_Policy.pdf).

49 http://www.sae.org/news/3550/.

50 See Tomotaka FUJITA, "On Featuring Automated Driving", 1501 *Jurist*, 14 (2017).

addition, § 3 of the Japanese Automobile Accident Compensation Act introduced liability for the operators of automobiles. The burden of proof for negligence was shifted to operators.[51] Here, the word "operator" is not limited to the driver but includes the owner of the vehicle (see § 2 (3) of the Act) as well as those who control the operation and gain benefits from the operation of the vehicle.[52] The Act also introduced a compulsory system of insurance for the aforementioned operator liability (§ 5). As a result, under § 3 operators shall buy insurance which covers the aforementioned liability for the operation of the automobile. Thus, victims are protected to a certain extent.[53] In addition, according to the Product Liability Act, manufacturers of automobiles can be held liable for "defects" in the automobiles they produce.

The aforementioned document *On the Legal Issues of the Automated Driving,* by the General Insurance Association of Japan, used the NHTSA framework of four levels of automation. It argued that regarding automation levels 1 to 3, the current framework of tort law and the Automobile Accident Compensation Act are applicable, whereas different provisions are necessary in respect of level 4. This is due to the fact that in level 4 automated vehicles, there is no natural human "driver" who is primarily liable either under current tort law or under the Automobile Accident Compensation Act, both of which assume the existence of a driver.

Tomotaka FUJITA[54] and Kyota KINNAI[55] have discussed similar liability issues under the Automobile Accident Compensation Act and argue that even in the case of fully automated vehicles, the Act is applicable and

---

51 That is, differently from normal tort liability according to § 709 of the Japanese Civil Code, according to § 3 of the Automobile Accident Compensation Act, the operator of an automobile can avoid liability for the accident caused by the automobile he or she operates only when he or she proves that the driver or operator exercised due care in operating the automobile.

52 See the Japanese Supreme Court Judgment of September 24, 1968.

53 It should be noted that there is discussion on whether the levels of compensation covered by compulsory insurance are adequate. What happens is that the compensation from the compulsory insurance is much smaller than the amount the court would award and therefore, people "hope" to be run over by rich drivers, because they (or their private insurance policies) will pay much more than what the compulsory insurance would pay.

54 Tomotaka FUJITA, "Automated Driving and the Operator's Responsibility", 1501 *Jurist*, 24 (2017).

55 Kyota KINNNAI, "Responsibility of Car Accidents by the Automated Vehicles" (2), 44 (11) *International Commercial Law* (kokusai-shoji-homu), 1612 (2016).

the operator can be held liable, because the user of an automated vehicle, when instructing an AI chauffeur to bring him to his destination, acts analogously to a person who asks a natural human chauffeur to bring him to his destination.

Atsumi KUBOTA[56] discusses sellers' and manufacturers' liability. The issues looked at involve product liability for automated vehicles, such as difficulties in the concept of "defect" or finding such a "defect" in actual accident cases.

Soichiro KOZUKA[57] discusses the importance of data protection in automated vehicles and argues in favor of privacy protection in respect to information collected by vehicles during automated driving.

III. Discussions on Automated Vehicles and the Criminal Law

In Japan, criminal liability resulting from car accidents is mainly governed by two sections in two statutes. One is the Act Concerning Punishment of Automobile Driving Causing Death or Injury *(jidosha unten shisho koi shobatsu ho),* i.e. through negligent driving, enacted in 2013 (Act 86 / 2013). § 5 of the 2013 Act stipulates that a person who fails to take the necessary care while driving an automobile and thereby causes the death or injury of another shall be punished by imprisonment with or without work for not more than 7 years or a fine of not more than 1,000,000 yen. The other is the former § 211 (1) Japanese Criminal Code (*Kaihoo* – JCC) (Causing Death or Injury through Negligence in the Pursuit of Social Activities) which stipulates that a person who fails to exercise due care required in the pursuit of social activities and thereby causes the death or injury of another shall be punished by imprisonment with or without work for not more than 5 years or a fine of not more than 1,000,000 yen. Traditionally, only § 211 JCC existed and Japanese courts had applied the provision to car accident cases for many years. In 2007, § 211 (2) was added to JCC § 211. It provided the same contents as in the current § 5 of the 2013 Act.[58] In 2013, the Act was passed and JCC § 211 (2) was moved to

56 Atsumi KUBOTA, "The Automated Driving and the Responsibility of Sales Companies and Manufacturers", 1501 *Jurist*, 30 (2017).

57 Soichiro KOZUKA, "Civil Liability and the Software Centered Automobiles", 1501 *Jurist*, 38 (2017).

58 Masahide MAEDA (ed.), *Commentary of Criminal Law*, 613 (3rd ed. 2013).

§ 5 of the new Act.[59] Now, in the case of car accidents, when drivers' criminal liability is at issue, § 5 of the Act is applicable and when other people's liability (such as the criminal product liability) is at issue, JCC § 211 is applicable.

Yoshihiko IKEDA[60] argues that at NHTSA levels 1 and 2, the current criminal law is applicable. As for level 3, IKEDA argues that drivers can be criminally liable for their involvement at the time of an accident or emergency, but as driver involvement is limited, the actual risk of liability is dramatically decreased. Regarding level 4, IKEDA argues that criminal liability is shared by the designers and manufacturers of vehicles.

Yuka NAKAGAWA[61] argues that if computer systems only assist drivers, there are virtually no changes in civil liability. When society fully trusts autonomous systems' safety, the rule of reliance on reasonable conduct (*Vertrauensgrundsatz*) will become applicable. NAKAGAWA also argues that in cases of automation similar to level 2, the question is whether the possibility exists of avoiding the result (*Vermeidbarkeit*) of death or serious injury to a victim when the driver is required to be involved in driving by, for example, an acoustic or visual warning sign. In the case of fully automated vehicles, NAKAGAWA also argues that drivers are not liable because they have no duty of care (*Sorgfaltspflicht*).

Masamitsu TOMIKAWA[62] points out the difficulty of allocating liability when automated vehicles are involved in accidents. This is because there are many sources of risk such as the hardware, software, misuse by the user, self-learning of autonomous machines, or a combination of some or all of these elements. This will lead to difficulty in determining and proving causation. Although Dr. TOMIKAWA argues that the drivers'/users' criminal liability would decrease dramatically because the expectation of a high level of safety in automated systems would likely negate foreseeability, Dr. TOMIKAWA acknowledges the existence of foreseeability in the case of manufacturers. As Dr. TOMIKAWA affirms the utili-

---

59 Atsushi YAMAGUCHI, *Criminal Law*, 226 (3rd ed, 2015).

60 Yoshihiko IKEDA, "The Relationship Between Criminal Law and Automated Driving System", 69(12) *Journal of Society of Automotive Engineers of Japan*, 33(2015).

61 Yuka NAKAGAWA, "The Transformation of Legal Liability for Traffic Accidents after the Introduction of Automated Driving System: The Difference between the Positions of Civil and Criminal Liabilities", 25 *Chukyo Lawyer*, 41 (2016).

62 Masamitsu TOMIKAWA, *loc. cit.* 2016 (f. 6) para 132.

ty of introducing automated driving systems, he recommends application of the concept of accepted risk (*Erlaubtes Risiko)* to limit manufacturers' liability.

D. *A Brief Analysis of Automated Vehicles Law in Japan*

I. The Difficulty of Allocating Liability because of Multiple Parties Involved

As some scholars have already noted, in the case of (partially) automated vehicles, the allocation of liability is much more difficult than in traditional automobiles.

Firstly, there are many parties involved in automated vehicles such as designers, manufacturers, sellers, ITS (intelligent transport systems) providers, telecommunication companies, etc. As a complicated system, it will be difficult to determine the real causes of accidents.[63]

Secondly, the autonomy, or the emergence[64] of automated vehicles also makes the issues more complex. Even the same models of automated vehicles will likely behave very differently within very short periods of time because technology and software will change.[65] As a result, causation will become more difficult to prove. The autonomy and the autonomous "judgment" of vehicles will make it difficult to determine the reasons why a certain decision was made when an accident occurs (wrong program, wrong information received at the time during programming or the wrong information acquired through the process of machine learning, etc.). Furthermore, autonomy will make it very difficult to predict the concrete be-

63 See Masamitsu TOMIKAWA, *loc. cit.* 2016 (f. 6).

64 There is a disagreement on the appropriate terminology between "autonomy" and "emergence". See Ryan Calo, "Robotics and the Lessons of Cyberlaw", 103 *Califfornia Law Review* 513, 538 (2015) and Jack M. Balkin, "The Path of Robotics Law", 6 *California Law Review* Circuit 45, 51 (2015). The author would use autonomy because in Japan this wording is more widely used.

65 "Consider the same model of AI vehicle we are planning to buy next Christmas: By gaining knowledge or skills from their own interactions with the living beings inhabiting the surrounding environment, the same model of AI chauffeur will behave quite differently after only a few days or weeks." Ugo Pagallo, *Law of Robots*, 72 (2013).

havior of automated vehicles and therefore, the predictability of road behavior in general will be reduced.

For these reasons, it is necessary to focus on how to better allocate liability in cases of accidents involving automated vehicles.

II. The Necessity of Striking a Balance between Over-Attribution and Under-Attribution

When considering the legal liability of vehicles with higher levels of automation, attention should be paid to the so-called *control dilemma.*[66] This has been discussed in a very meaningful way by Professor Eric Hilgendorf. Even as technology enables less and less monitoring and control by drivers, the law still requires drivers to continuously monitor and control their vehicles. This situation not only decreases the attractiveness of automated vehicles but also forces drivers to attempt to pay attention although an adequate level of stimuli is not present.[67]

This control dilemma raises a further question: how to strike a balance between over-attribution and under-attribution when determining the legal liability of parties involved in accidents of automated vehicles. That is, the law should strike a balance between the alarming effects of automation by imposing neither too much liability (over-attribution), nor too little liability, and thereby running the risk of not protecting victims (under-attribution).

If, for example, the liability allocated to users is too high, then users will need to take actions to protect themselves. A user might, for example, pay increased attention to his surrounding environment and monitor the automated driving system of the vehicle in order to avoid incurring liability even after full automation becomes technically possible. Some users

66 Eric Hilgendorf, "Automatisiertes Fahren und Recht", *Veröffentlichung der auf dem 53. Deutschen Verkehrsgerichtstag vom 28. bis 30. Januar 2015 in Goslar gehaltenen Vorträge, Referate und erarbeiteten Empfehlungen*, Köln 2015, 67.

67 In Prof. Hilgendorf's words: "*Control dilemma* is the term used to refer to a phenomenon which arises in the context of autonomous vehicles: technological development allows drivers to be increasingly relieved of monitoring and control tasks, while the law continues to require drivers to substantially drive and control the vehicle. This not only considerably reduces the attractiveness of new cars (and trucks), but also leads to driver overload, as psychological tests show that sustained attention is not possible without sufficient stimulation."

might even decide not to use the automated system for fear that their risk of incurring liability is too high. The same is applicable to the designers, programmers, manufacturers or other parties involved in the production of automated vehicles and systems. They might avoid designing/programming/manufacturing automated vehicles for fear of heightened levels of liability. Considering the aforementioned advantages of fully automated vehicles, it is not expedient for the law to inhibit the development and use of automated vehicles. Therefore, the law should not become too rigid or attribute too much liability to the new technology (i.e., automated vehicles) in order to avoid creating disincentives.

On the other hand, when regulation is too loose, and most of the parties involved are likely to avoid liability if accidents occur, then accident victims are left unprotected. Such a situation would cause social instability, because regardless of how wide the use of automated vehicles becomes, it will not be possible to reduce the number of car accidents to zero. There will always be a number of car accident victims who suffer death or serious injury. Therefore, it is essential that the rules on liability should not be too lenient.

What is necessary is to strike a balance between over-attribution and under-attribution. The legislature should enact laws which prescribe adequate levels of liability for all parties. Towards this end, legal scholars have discussed many "legal tools". The following discussion is a preliminary analysis of the potential use of such tools in striking such a balance both in civil law and criminal law contexts.

III. Tools to Strike a Balance in the Field of Civil Law

There are many kinds of tools available in civil law, such as foreseeability, ways of avoiding damage, causation, developmental risk defense (in PL laws), safe harbor rules, burden of proof, strict liability, compulsory insurance,[68] or the acknowledgment (to some degree) of the legal personality of automated vehicles as such.[69]

---

68 See Gerald Spindler, "Roboter, Automation, künstliche Intelligenz, selbststeuernde Kfz – Braucht das Recht neue Haftungskategorien?", 2015/12 *Computer und Recht*, 766 (2015).

69 Such as the digital peculium proposed by Ugo Pagallo (Ugo Pagallo, *Law of Robots*, 103 (2013).).

Regarding the issue of under-attribution, it should be noted that although uncertainties still exist,[70] there are scholars who argue that the current compulsory insurance system is applicable to fully automated vehicles.[71] At least such compulsory insurance can be a good mechanism for resolving the issue of under-attribution because victims can obtain compensation in damages from insurance companies for the injuries they have sustained.[72]

The next question would be how to resolve the issue of over-attribution. Under the assumption that the current compulsory insurance system would still be applicable to fully automated vehicles, the question regarding over-attribution would be why users (operators) should still buy (potentially expensive) compulsory insurance when they are unlikely to be negligent in operating fully automated vehicles. So it is one idea that other parties such as car manufacturers or ITS service providers should buy and carry compulsory insurance. However, this creates a new problem: manufacturers' costs (or any other parties' costs) will ultimately be shifted to the users/buyers in the form of increased purchase prices for automated vehicles. Thus the current compulsory insurance system of forcing operators/users to buy insurance might remain the best solution after all.

IV. Tools to Strike a Balance in the Field of Criminal Law

Tools available in civil law are also available for striking a balance in criminal law. These include foreseeability, avoidance of damage, causation, the rule of reliance on reasonable conduct, assumption of risk (*Risikoübernahme*), accepted risk, safe harbor rules, or the acknowledgment (to some degree) of the legal personality/criminal liability of automated vehicles.

In the context of criminal law, the tools of foreseeability and the avoidance of damage might not work very well, because it is always foreseeable

70 See e.g., "On the Legal Issues of the Automated Driving": http://www.sonpo.or.jp/news/file/jidou_houkoku.pdf.

71 Tomotaka FUJITA, *loc. cit.* 2017 (f. 51), Kyota KINNNAI, *loc. cit.* 2016 (f. 52).

72 It should be noted, however, that under the Automobile Accident Compensation Act, although the burden of proof is on the operator (user), it is theoretically possible for the user to avoid liability by proving that the user was not negligent and in such case, the victims should sue other parties and prove negligence or the existence of the defect.

in the *abstract* that automated vehicles will cause accidents. *Theoretically* it is possible to avoid this risk by not designing, manufacturing, or using automated vehicles. Thus, the concept of avoidance could be applied arbitrarily by judges (at least it is unpredictable how judges would decide) and as a result, the foreseeability and the possibility of avoiding the result would not work well as tools.

As for reliance on reasonable conduct and assumption of risk, these tools might work well in some contexts but not in others. Reliance on reasonable conduct can be used when autonomous vehicles rely on human victims' compliance with traffic rules.[73] But it is not clear to what extent this tool can be extended to other issues. For example, can a person who relied on the proper operation of a (partially) automated vehicle plead innocence because his reliance is justified[74]? Furthermore, assumption of risk can be used to protect manufacturers in cases of accidents where users who bought automated vehicles are the victims. The argument is that they decided to use the vehicle and take the risk, i.e. they gave informed consent. However, where a third party is injured in an accident, the third party could not have assumed the risk, so this concept does not work in that situation.[75]

Accepted risk is a relatively useful tool.·[76] This concept seeks to balance the risk of causing harmful results with the potential benefits of using such technology. In crimes of negligence no criminal liability is incurred when the damage falls within the scope of risk which is allowed.[77] As technology is understood to bring benefits to society, the concept of accepted risk would work well for automated vehicles. Using this concept, judges could strike a balance between over-attribution and under-attribution when they decide whether criminal liability has been incurred in concrete cases under § 5 Act 86 / 2013 Act (Act Concerning Punishment of Automobile Driving Causing Death or Injury) or § 211 JCC (Injury through Negligence in the Pursuit of Social Activities). However, this concept might not work well because when accepted risk is applied by judges in individual cases,

---

73 In the Supreme Court Judgment December of 20, 1966, the human driver was acquitted because he relied on the victim's compliance with traffic rules. This rule should also be applicable in the case of the automated vehicles.

74 See Yuka NAKAGAWA, *loc. cit.* 2016 (f. 58).

75 See e.g., Jan Philipp Günter, *Roboter und rechtliche Verantwortung*, 224 (2016).

76 Eric Hilgendorf, *op. cit.* 2015 (f. 63), para 64.

77 Atsushi YAMAGUCHI, *op. cit.* 2016 (f. 56), para 249.

they consider all the circumstances. Designers, manufacturers, and users, in contrast, do not know beforehand whether their conduct (i.e., the risks created) will be considered accepted risks.

From a legal policy or social welfare perspective, if the decision is made to promote the development and use of automated vehicles, it might be better for the legislature to amend the law and provide clear safe harbor rules. For example, a new law could be enacted stipulating that designers, manufacturers and sellers shall not be liable in negligence for accidents involving their automated vehicles where registration of the vehicle was approved by the responsible government authority or the vehicle conformed to certain technical standards. Alternatively, a law might be enacted stipulating that users of such automated vehicles, which have obtained registration approval or conform to the appropriate technical standards, would have liability for misuse or for failure to adequately maintain their vehicles. These clear rules would significantly reduce disincentives to manufacturers, etc. arising from potential criminal liability for accidents.

Finally, if accepted risk or safe harbor rules were used, there would be more situations where no one is criminally liable for tragic accidents. Such under-attribution might face significant opposition. One approach would be that such situations can be justified as long as victims are fairly compensated for their injuries, for example, by compulsory civil liability insurance. But if under-attribution is really problematic in terms of the criminal law, we should consider the introduction of a digital person concept in which a robot or automated vehicle itself could incur criminal liability.[78] Dr. Yoshiaki ITO, along with other scholars,[79] maintains that such a scenario strongly argues against criminal liability for robots because punishing robots necessarily entails placing a burden or inconvenience onto natural persons such as their owners. The issue arises as to whether such natural persons would be indirectly punished if "criminal liability of robots" existed even though these natural persons were not negligent nor did they cause anyone injury.[80]

78 See Eric Hilgendorf, "Können Roboter schuldhaft handeln?" in Susanne Beck (ed.), *Jenseits von Mensch und Maschine* (2012), 119.

79 See, e.g., Jan C. Joerden, „Strafrechtliche Perspektiven der Robotik" in Eric Hilgendorf & Jan-Philipp Günter (eds.), *Robotik und Gesetzgebung* (2013), 195.

80 Yoshiaki ITO, *loc. cit.* 2016 (f. 6).

Automated Driving and the Law

*Prof. Dr. Dr. Eric Hilgendorf,**
Universität Würzburg

A. Terminology

Attention should be paid, first of all, to our use of language: whilst use is frequently made in the mass media of terms like "robot-car" or "driverless car", expressions such as "assisted driving", "smart driving" or "piloted driving" are preferred by manufacturers and in the specialist literature.[1] In what could be called basic research or engineering oriented literature, use of the terms "autonomous driving" or "partially autonomous driving" seems to be most widespread. The German Federal Highway Research Institute (BASt) eventually proposed use of the term "automated driving", suggesting that various different levels of automation be distinguished: "driver only", "assisted", "semi-automated", "highly automated" and "fully automated ".[2]

Terminology issues are important if one wants to be able to express oneself clearly and precisely. On the other hand, technological development is moving so quickly that it would seem inappropriate, from the outset, to attempt to use definitions to force developments into certain categories. On closer inspection, objections can be raised to all of the levels proposed in the BASt report.

One objection to use of the term "automated driving" is the fact that since the time when horse-drawn carriages became obsolete, almost all ve-

---

* The original text was written for the 54th German Traffic Court Conference (Verkehrsgerichtstag) in Goslar on February 25th to 27th, 2016. For publication the original text was retained, but updated and supplemented by footnotes.

1 Cf. above all Köhler & Wollschläger, *Die digitale Transformation des Automobils*, 2014; Laglstorfer, *Die Zukunft des intelligenten Automobils*, 2013; Meyer & Beiker (eds.), *Road Vehicle Automation*, 2014; Rubinyi (ed.), *The Car in 2035. Mobility Planning For the Near Future*, 2013; Siebenpfeiffer (ed.), *Vernetztes Automobil*, 2014.

2 Gasser *et. al.*, *Rechtsfolgen zunehmender Fahrzeugautomatisierung. Gemeinsamer Schlussbericht der Projektgruppe*, 2012, p. 9 (BASt report, pamphlet F 83).

hicles on the road have been automated vehicles. Use of the term "automated automobile" therefore appears to be pleonastic. An "autonomous system" is a kind of technical system that can cope with problems intelligently in various situations, without having to rely on human input.[3] In principle this covers well the relevant legal characteristics of the systems under discussion here, such as lane departure prevention systems or traffic jam assistants.[4] What is problematic with the use of the terms "autonomous" or "partly autonomous", however, is their allusion to questions of artificial intelligence on the one hand, and classical German philosophy on the other. Historically, "autonomous" was used, for example by the Greeks, to describe a people or place which was able to govern itself or legislate for itself. In philosophical language it meant a person who was capable of giving laws to himself.[5] This traditional understanding of the term does not fit easily as a way of describing today's cars. Similarly, terms like "piloted"[6] or "assisted driving" have advantages but also have disadvantages.

In this contribution, despite some reservations, I will follow the German Government's usage of the term "automated driving". In order to avoid any misunderstanding, however, one point has to be made clear here: definitions are in themselves not able to resolve legal questions. It is on no account possible to derive legal obligations (duties) or other legal consequences from a particular definition of automated driving or level of automation without reference to the relevant legislation. It is only when the legislature attaches legal consequences to a particular definition that the definition becomes legally relevant.

---

3 Häußling, *Techniksoziologie*, 2014, p. 89.

4 For an overview, cf.."Das Auto von Morgen. Wie futuristische Technik unser Leben verändert" in *Der Spiegel Wissen*, Nr. 4, 2014; cf. also the articles on "assistance systems" in *Automobiltechnische Zeitschrift* (ATZ) 2014, issue 02, p. 10 *et seq.*, as well as articles on the topic of "automation and safety" in *Automobiltechnische Zeitschrift* 2015, issue 01, p. 14 *et seq.*; cf. also articles on "trends in technology in commercial vehicles" in *Automobiltechnische Zeitschrift* 2014, issue 09, p. 10 *et seq.*

5 Such a definition of the concept of "autonomy" is, as a rule, traced back to the philosopher Kant.

6 Use of the expression "piloted driving" reminds us of parallels to air transport, where, of course, autonomous systems have been used for many years, remarkably, without any doubt being expressed as to whether such technology can be licensed for use.

B. Convergence of Technologies

Progress in automated driving can hardly be clearly distinguished from advances in other areas of technology. This includes, in particular, areas such as robotics or autonomics, sensor technology, and lastly, advances in information technology and communication technology[7] in general. Conclusions drawn from legal analyses and assessments in these fields can therefore often be applied *mutatis mutandis* to the problems of automated driving.[8]

The convergence of technologies corresponds to a convergence of themes: the debate on autonomous systems and general legal issues associated with them includes legal problems associated with automated vehicles. In the same context we also find buzzwords like "Industry 4.0", "Smart City"[9] and "Smart Homc" as well as the latest development, exploring the possibilities for a "predictive user experience". This refers to attempts to predict human needs and human behaviour in such a way that the computer becomes like a caring and attentive friend.[10]

The central characteristics of this impending transformation in the technology of road transport can be subsumed under four terms: *automation* (largely independent of human input), *interconnectedness* (barrier-free exchange of information throughout the network), *adjutory* (helpful – the technical system knows preferences of the user and addresses them of its own initiative), and *ubiquitous* (the appropriate technology is available – almost – everywhere). Road transport will therefore become part of the "Internet of Things".[11]

---

7 Hilgendorf, "Recht und autonome Maschinen – ein Problemaufriss" in Hilgendorf & Hötitzsch (eds.), *Das Recht vor den Herausforderungen der modernen Technik. Beiträge der 1. Würzburger Tagung zum Technikrecht im November 2013*, 2015, p. 14 *et seq*.

8 This applies, for example, to research on the interdisciplinary topic of "law and functional safety in autonomics" (*Recht und funktionale Sicherheit in der Autonomik*), which was examined by the project "Autonomics" (Autonomik) conducted under the auspices of the German Federal Ministry of Science between 2010 and 2013, see www.autonomik.de.

9 For a comprehensive treatment, see Kaczorowski, *Die smarte Stadt – Den digitalen Wandel intelligent gestalten*, 2014.

10 For more depth, cf. http://next.mercedes-benz.com/pux-de. See also Kuri, "Technik denkt mit. Mit dem autonomen Auto ins intelligente Haus" in *c't Magazin für Computertechnik* 2015, issue 4, p. 92 *et seq*.

11 Hilgendorf, *op. cit.* 2015, p. 15.

C. The Advantages and Disadvantages of Automated Road Transport

Before examining the legal problems, and in particular the problems in constitutional law of developing automated vehicles, it would be sensible to glance briefly at the advantages and disadvantages these new vehicles might offer.

The first advantage that needs to be mentioned is increased road safety: Experts believe that up to 90% of road accidents may be attributed to human error.[12] The increased use of autonomous systems in cars would probably reduce the number of fatalities per year by a significant number. Another advantage of automated vehicles would be the possibility of increasing the mobility of older persons or handicapped persons, an issue that is often neglected in discussions. Demographic change will lead to a situation where more and more elderly people want to continue driving in their older years. Autonomous systems are extremely well suited to mitigating or completely compensating for age-related handicaps, thus allowing these people the possibility of retaining a high level of mobility. It can even be expected that people can become mobile again, by means of these new vehicles, who were previously housebound.[13]

Additional points which speak in favour of automated driving are improved environmental protection, the possibility of a rational use of resources, i.e. primarily lower energy consumption, the promotion of ease of transport, especially in urban areas, increased convenience and freedom when driving, time savings, and finally the possibility of new business models, for example in the form of car-sharing concepts. Particularly the last point shows that automated driving also holds enormous commercial potential.[14]

On the negative side, one would tend to think first about the cost of the new technology, which, however, according to all previous predictions,

---

12 Wachenfeld & Winner, "Der Sicherheitsnachweis für autonome Fahrzeuge" in Hilgendorf, Hötitzsch & Lutz (eds.), *Rechtliche Aspekte der Fahrzeugautomatisierung. Beiträge der 2. Würzburger Tagung zum Technikrecht im Oktober 2014*, 2015, p. 53. The authors noted that depending on the source of the data, there was scattering in the numbers.

13 The idea of establishing the Würzburg RobotRecht Research Centre (www.robotrecht.de) first arose during a symposium on the legal problems of automated, self-propelled wheelchairs.

14 For more information, see Laglstorfer *op.cit.* 2013; for a comprehensive treatment see Proff (ed.), *Radikale Innovationen in der Mobilität*, 2014.

will stay within acceptable limits for the consumer. Once such properly equipped vehicles go into mass production, the additional costs will hardly be relevant any more. There is a much more serious point, which could be called "disenfranchising the driver": the use of automated systems means that the driver is asked to make fewer and fewer decisions. This is certainly a serious point in a country like Germany, where driving is also, and perhaps above all, associated with "leisure activities" and "sports", where "fun driving" or "driving pleasure" are important. Autonomous vehicles could therefore be perceived by many drivers as “killjoys”. Of course engineers will be able to take account of and indeed partly resolve this problem by designing the new systems in such a way that they can at least be partly overridden, so that it remains possible to switch them back to the “classic” sport-driving mode (outside of rush hour, of course).

Another point which should be considered at an early stage is potential pressure to adapt to technical legal requirements. In a humane legal system, technology serves mankind and not the other way around.[15] A further problem is the risk of dependence on monopoly suppliers, in particular with regard to the software used. It is exactly because road transport will become part of the Internet of Things that the risk exists that US providers such as Google, Facebook and others, who already dominate the Internet, will also dominate (or try to dominate) communication within and between autonomous vehicles, so that they can unilaterally set standards in this area. Such a development definitely would not only be a double edged sword in commercial terms, but also with regard to the interests of consumers.[16]

In my opinion the biggest problem connected to the introduction of automated cars currently is the significantly increased susceptibility of the new vehicles to sabotage. The car of the future will not be a stand-alone vehicle, but rather when it is underway, it will be networked so that it is getting information from the Cloud, about events on the road, for example, but also about the weather or obtaining the general news. The vehicles will

---

15 Hilgendorf, “Humanismus und Recht –Humanistisches Recht? Eine erste Orientierung“ in Horst Groschopp (ed.), *Humanismus und Humanisierung*, 2014, p. 36 *et seq*. (55).

16 The existence of several large, competing firms generally has favourable effects for the consumer. This changes when the firm acquires a monopoly. German law provides a range of instruments which should prevent or preclude this, starting with antitrust law and going all the way to the possibility of expropriation.

also communicate with each other via the infrastructure of the road, or, respectively, of the "smart city" in which they are travelling. One need only mention communication between cars and trucks. These communications networks are apparently extremely interesting for hackers. This means that events on the road will be at risk of being exploited in some way by cybercriminals. The problems associated with this are enormous.[17]

D. Constitutional Provisions

The most important legal issues being discussed today in relation to automated driving have to do with the certification for road use and running of automated vehicles. This includes issues of civil liability, criminal liability, provider liability, individual privacy protection, and lastly, insurance law. Constitutional principles, in a certain sense, take priority over all of the other issues just mentioned. As this topic has hardly been touched upon in the literature up to now, I will allow myself a number of remarks here.

New technologies are hardly mentioned in the German Federal Constitution (GG), at least as far as the actual wording is concerned. Nevertheless, it is beyond dispute that the freedom to conduct research contained in Art. 5(3) GG includes the right to develop new vehicle technologies. Production and distribution are protected by Art. 12 and Art. 14 GG as well as other provisions. Thus the law applicable to automated vehicles is exactly the same law as applies to other technologies.[18]

According to the jurisprudence of the German Constitutional Court (*Bundesverfassungsgericht*), fundamental rights in their totality form an "objective value system", which not only serves as an aid in interpretation,

---

17 In any case the topic of "cyber-security" has drawn increased attention in recent years. A comprehensive collection of articles may be found on the web pages of the German Federal Office for Information Security (BSI), https://www.bsi.bund.de. For a spectacular break-in into a BMW ConnectedDrive vehicle via SMS, cf. paar, "Auto, öffne dich!" in *c't Magazin für Computertechnik* 2015, issue 5, p. 86 *et seq*.

18 For further information, cf. Hilgendorf, "Teilautonome Fahrzeuge und das Recht. Verfassungsrechtliche Vorgaben und rechtspolitische Herausforderungen" in Hilgendorf, Hötitzsch & Lutz (eds.), *op.cit.* 2015, p. 15 *et seq*.

but also creates obligations for the State.[19] Thus, the State is obliged to protect the physical integrity and the lives of its citizens. From this it can be derived that the State is required to promote those technologies that contribute demonstrably to the protection of constitutionally protected legal rights or interests, and at the same time at least not to unnecessarily hinder that development. This includes a technology-friendly interpretation of relevant legal provisions, provided that the technologies in question are capable of protecting the above-mentioned constitutional rights – which seems to be uncontroversial in the case of automated driving.

In Germany, the Principle of Human Dignity is often used in arguments against new technologies. It would be a mistake to trivialize Art. 1 GG by (mis-)using the guarantee of human dignity as a cloak to cover-up conservative research policies and industrial policies. The guarantee of Human Dignity should rather be interpreted very narrowly so as not to be devalued by inflationary use into "small change". If one accepts this premise, then no objections to automated driving can be derived from the Human Dignity guarantee. On the contrary, the increase in mobility for the elderly and the disabled would be evaluated very positively from the perspective of Human Dignity.

The question is, what conclusions can be derived for automated driving from the "fundamental right to freedom of movement".[20] It does not need to be decided here whether this is a separate fundamental right in German law; it is undisputed that freedom of movement on the roads enjoys constitutional legal protection. Infringements of this right, therefore, require legitimate reasons and must be suitable and necessary to achieve their legislative objectives. One could argue that the introduction of automated vehicles onto the roads would involve disadvantages for the drivers of non-automated vehicles, so that their constitutionally protected rights might be affected. In the absence of detailed evidence, however, this argument seems unconvincing, especially since the introduction of highly automated vehicles is beginning slowly.

---

19 This has been established case law since the German Constitutional Court judgment in *Lüth*, see BVerfGE 7, 198 *et seq.*; for a comprehensive treatment, cf. Dietlein, *Die Lehre von den grundrechtlichen Schutzpflichten*, 2nd ed., 2005.

20 Ronellenfitsch, "Die Verkehrsmobilität als Grund- und Menschenrecht. Betrachtungen zur 'zirkulären' Mobilität in der Europäischen Union", in *Jahrbuch des öffentlichen Rechts der Gegenwart*, vol. 44 (1996), p. 167 *et seq.* (p. 182 *et seq.*).

It is occasionally argued that automated driving might violate the principle of criminal culpability (fault-based liability).[21] This argument, however, appears rather far-fetched.

The principle of culpability states that the culpability of the offender is the basis for all punishment and that, moreover, the penalty must be restricted to what is appropriate considering the extent of guilt of the offender.[22] As it is stated here, the principle of culpability reflects the constitutional principle of rule of law. It would violate the principle of culpability to impose punishments on machines, but up to now this has not been called for seriously. The introduction of automated systems in cars could lead to a situation where the number of crimes clearly attributable to human beings would be reduced; that, however, would certainly not be a violation of the culpability principle.[23]

Taken all together, what is important is that our reference to the constitution has not indicated any incompatibility which would speak against the legal certification of autonomous vehicles for use on German roads. In view of the significant advantages which the new technologies promise, one would even have to assume that the State had an obligation to promote the new technologies. If no changes were made, this could still be accomplished through a technology-friendly construction of the current applicable law. But one should also consider what the law should be like in the future; the legislature should not hinder sensible technological developments but rather it should support them, and under certain circumstances might even be under a constitutional duty to do so. A good legal framework could even be a competitive advantage in international competition, if it were possible to get world consumers to associate the ideas of a consumer-friendly, environmentally friendly, or particularly "humane" legal framework to regulate technology with the designation of origin “Made in Germany”.

---

21 This position was expressed, for example, by a representative of the German Ministry of Justice at a meeting of the “round table“ at the German Ministry of Transport on 26 February 2015.

22 BVerfGE 9, 167 (169), 95, 96 (131); BGHSt 2, 194 (200); Hilgendorf & Valerius, *Strafrecht Allgemeiner Teil*, 2013, § 6 paragraph 1.

23 Otherwise any other measure for the avoidance or prevention of crime would be contrary to the principle of culpability, an obviously absurd idea.

E. Legal Issues of Licensing and Vehicle Operation

The licensing for use and operation of automated vehicles is still very problematic in law. Until recently, the Vienna Convention on Road Traffic of 1968[24] was considered to be a significant hurdle for automated driving. In point of fact, the drafting of the Convention, particularly Articles 8 and 15, can be construed to support the view that vehicles may only be licensed for road use which are driven by a human driver. Moreover, the convention makes use of the phrase "control of the driver" in Articles 8 and 13, which may also be used as an argument against the legality of automated driving.[25]

In any case, in early 2014 amendments to the Vienna Convention were agreed which would make it lawful to license automated vehicles with levels of automation up to the level "high automation".[26] The extent to which German regulatory law is affected by these changes is not entirely clear: one should at least be able to assume now that there are no longer provisions in the current wording of the Vienna Convention which would stand in the way of amending the rules on driving contained in the German Highway Code which are relevant to automated driving.[27] One must of course bear in mind that the provisions of the Vienna Convention are not directly effective in Germany, but rather must be implemented into the domestic traffic code (and elsewhere into domestic law) for them to take effect. Until such changes are made, German road traffic law remains unaffected.

It would be a mistake to make predictions on how the law will be amended based on all the various competing definitions of automated driving currently making the rounds.

---

24 The current version can be found at https://www.ris.bka.gv.at/Bund.

25 For a summary, see Geissl, "Automatisiertes Fahren – Vereinbarkeit des technisch Machbaren mit geltendem Recht", in *Recht – Automobil – Wirtschaft. Sonderausgabe des Betriebs-Berater* 2013, p. 20 *et seq.*; Hötitzsch & May, "Rechtliche Problemfelder beim Einsatz automatisierter Systeme im Straßenverkehr" in Hilgendorf (ed.), *Robotik im Kontext von Recht und Moral*, 2014, p. 189 *et seq.*; 192 *et seq.*; Lutz, "Anforderungen an Fahrerassistenzsysteme nach überstaatlichem Recht" in Hilgendorf & Hötitzsch, *op.cit.* 2015, p. 171 *et seq.*

26 Lutz, "Autonome Fahrzeuge als rechtliche Herausforderung" in *Neue Juristische Wochenschrift* 2015, p. 119 *et seq.*, at p. 123.

27 For a similar opinion, cf. Lutz, *ibid.*

Thus, the definitions proposed by BASt referring to various different levels of automation, which were discussed at the round table of the Ministry of Transport, have the advantage of providing a unified vocabulary for future discussions.[28] Otherwise they obviously have no legal relevance and in particular do not in any way change the duties an automated vehicle would be subject to, namely to drive safely and carefully. The fact that such a misunderstanding could arise at all shows that the original language is to some extent flawed. Thus, the statement that the automated driver "must" drive in a certain way does not mean that a legal duty has been established; rather it means that some technical imperative exists. The same thing could be said differently: if a vehicle with XY level of automation is to be driven safely in traffic, then it must be driven in this or that way. There is no BASt workgroup tasked with establishing binding rules to be applied to driving in German road transport law or in German negligence / tort law. Accordingly, until the German traffic code is amended, drivers will in principle be prohibited from engaging in activities which could impair their driving or distract them from their driving.[29]

F. Civil Liability

With respect to the issue of automated driving, a second major problem area in the general law is civil liability in tort. The most important relevant provisions are contained in § 823 German Civil Code (*BGB*), the Product Liability Act (*Produkthaftungsgesetz*), and § 7 and § 18 of the Road Traffic Act (*Strassenverkehrsgesetz*). The strict liability provision contained in § 7 of the Road Traffic Act is independent of the level of automation of the vehicle. In the course of technological development, one would expect to see a shift in liability from the driver to the manufacturer of the automated systems. One might therefore want to rethink the civil liability limits contained in § 12 StVG. This topic is not the subject of my presenta-

28 See footnote 1. For an extensive treatment see Gasser, "Die Veränderung der Fahraufgabe durch Fahrerassistenzsysteme und kontinuierlich wirkende Fahrzeugautomatisierung" in *Deutsches Autorecht* 2015, p. 6 *et seq.*

29 On this subject, see also the relevant resolutions of the German Traffic Law Conference held in Goslar, 2015.

tion.[30] In any case, I do not see the need for any large-scale legislative action in this area.

G. The Criminal Law

In the area of criminal law, the need for legislative action is also quite restricted. Within the core of the German Criminal Code (StGB) itself, §§ 212, 222, 229 and § 315c StGB will need attention. As regards misdemeanours and administrative offenses, thought should be given to amending §§ 24 and 49 of the Traffic Code (StVO). There will be no need to create any new offences in the foreseeable future.[31] The most important issue that needs to be faced is how liability in negligence will be dealt with in the context of automated driving. Regarding this issue, the reader is warned again against drawing false or hasty conclusions, namely the temptation of drawing conclusions about the relevant legal obligations in individual cases from the definitions of automated driving currently in use in discussions on the subject.[32] What the law of negligence requires of an individual in a specific case is decided by the competent court.

This means that it will be technically possible in the near future to give at least partial control of the vehicle over to the automated driving system. Nevertheless, the driver will remain subject to the duty of care requirements contained in previous court jurisprudence. We can only assume that after automated vehicles have been driven largely error-free over a longer period of time, the duty of care requirements placed on the drivers will be gradually reduced by the courts.[33] Of course, what we are facing here is a dilemma about control of the vehicle. The law is lagging behind: it is technically possible for the human driver to give up control of the vehicle to the computer, but this option is not legally possible as the human driver

30 Cf. Lutz, *op.cit.* 2015, p. 119 *et seq.*

31 Nevertheless, one could argue that soon new kinds of punishable behaviour will be possible, such as sleeping during the operation of a vehicle. Whoever drives while severely intoxicated incurs criminal liability (§ 316 StGB), but as soon as he falls asleep, the applicability of § 316 will at the very least be doubtful (How can someone "drive a vehicle" while being unconscious?). Even politicians with very punitive attitudes would be well advised to wait for further developments.

32 See above, footnote 2. Lutz *op. cit.* 2015 is therefore ambiguous, see p. 121.

33 The development of automated driving in the vertical direction that is to say the use of elevators, was very similar.

has a duty to control the vehicle.[34] Naturally, this "control dilemma" could be mitigated by legislative action.

For a driver to be able to adapt the level of his driving input to the level required in automated driving mode, he has to be able to recognise from instant to instant the degree of automation provided by the vehicle. One problem which exists in practice is that a new vehicle does not simply drive itself at a specified level of automation. The automated driving system is actually composed of a number of subsystems, each of which may be functioning at a different level of automation. This is the case because, on the one hand, the subsystems were designed separately and then subsequently integrated into a single automated system, and, on the other hand, because the driver can override (i.e. turn off) individual subsystems. So, for example, it would be possible for the lighting system to be set on fully automatic, for the lane departure prevention system to be set at an intermediate level of automation, and for the parking assistant to be completely switched off because the driver is convinced he can always park the car better than the on-board parking assistant. This is also a reason why the BASt definitions, which give an across the board rating of the automation level of the vehicle,[35] are in a certain sense misleading.

In order for the driver to meet all the specific care requirements necessary to fulfil his duty of care, he has to pay attention to all the systems in the vehicle which contribute to the total risk of damage. In Germany, the duty of care is established using two variables: risk of potential damage and the possibility of avoiding the risk. The allowable risk ("erlaubtes Risiko") and the reliance principle ("Vertrauensprinzip"), an essential principle for road transport under which one may rely on other drivers to drive safely, limit the duty of care.[36]

A major challenge lies in criminal liability for the vulnerability of a system to sabotage.[37] If a criminal infiltrates the data processing system of a

34 Hilgendorf, "Autonome Autos. Wer haftet für Roboter?" at http://www.lto.de/recht/hintergruende/h/autonome-autos-google-car-haftung-verkehrsrecht.

35 See above, footnote 2.

36 On the determination of duty of care obligations, see Hilgendorf & Valerius, *op.cit.* 2013, § 12 paragraph 17 *et seq.*

37 This problem arises for the whole "Internet of Things", especially for networked factories ("Industrie 4.0"), and should today even be considered to be one of the most important legal challenges in the ICT sector. See also the report published by BSI, "Die Lage der IT-Sicherheit in Deutschland 2014", p. 10 *et seq.*, available at https://www.bsi.bund.de/DE/Publikationen/Lageberichte/bsi-lageberichte.html.

vehicle and causes an accident, it is not just the hacker who should incur criminal liability. One should also consider whether criminal liability ought to be incurred by the person or persons who designed and installed the security system into the vehicle, as the failure to install an adequate security system could also be seen as a causal factor for the resulting accident. As the "Internet of Things" is gradually enlarged, there will be a concomitant increase in the risk of sabotage by IT attacks. The risks in road transport could be enormous; the danger exists that road transport in the "Smart City" of the future could in fact become a golden opportunity for Internet criminals.

Up to now, issues related to these developments have hardly been discussed; also here the concept of allowable risk should be given a decisive role in tackling new liability issues. Moreover, road transport issues need to be addressed as soon as possible and integrated into the IT Security Bill which is currently under consultation.[38] Finally, there is a lot to be said for amending the liability regime for internet providers in order to ensure their support and continued cooperation in the fight against internet criminality.

H. Provider Liability

The liability regime of Internet providers becomes particularly important if one considers the high level of interconnectedness necessary for automated vehicles to function. Providers are the firms that grant access to the Internet for data traffic. The German Telemedia Act (TMG) distinguishes between various kinds of providers in §§ 7 – 9: access providers provide a portal for entry onto the Internet, hosting providers make space available for others to store content on the Internet and finally, content providers publish content on the Internet. In particular, the law currently gives host and access providers not insignificant privileges[39] (i.e. immunity from prosecution). This raises the question of whether such privileges can continue to be tolerated and granted, given that in road transport the risk of injury will be multiplied many times over. Regarding the impact of cyber-

38 http://www.bmi.bund.de/SharedDocs/Downloads/DE/Nachrichten/Kurzmeldungen/entwurf-it-sicherheitsgesetz.pdf.

39 On the system of provider liability, see Hilgendorf & Valerius, *Computer- und Internetstrafrecht. Ein Grundriss*, 2nd. ed. 2012, paragraph 193 *et seq*.

crime, road transport represents a new dimension, not only in terms of the number of possible victims who may be attacked, but also in terms of the severity of the damage, i.e. to life, limb and property, which may be caused (this could be potentially comparable to networked industrial plants, as has been under discussion under the catch phrase "Industry 4.0"). It seems to me that it is no longer acceptable or reasonable that access providers who feign deafness in response to warnings about the new risks rather than withdrawing Internet access from cybercriminals should continue to get blanket exemptions from criminal liability.[40] But beyond that, we need to check whether and to what extent it makes sense to give Internet providers an incentive to contribute to road safety by making them liable for the criminal actions of others which are only possible because of the Internet access they provide.[41]

I. Data Protection and Privacy Law

Data protection and privacy will be of great importance in the road transport of the future. It will soon be possible to view private automobiles, without exaggeration, as "mobile data collection points". The problem was discussed at the 2014 German Traffic Law Conference[42] and is outside the scope of this paper. Please allow me, however, to touch upon several important aspects: under the current German law, data sovereignty for the data collected by a motor vehicle is in principle the responsibility of the keeper or owner of the vehicle as the data at issue are personal data or at least person-related data.[43] In light of the fact that today, personal data can be generated from almost any data, the question arises as to whether the

---

40 For an overview of the state of opinion on this subject, see *ibid.* paragraph 215 *et seq.*

41 On this subject, see also Kian & Tettenborn, "Ist die Providerhaftung im Lichte vernetzter autonomer Systeme noch zeitgemäß?", in Hilgendorf, Hötitzsch & Lutz (eds.) *op. cit.* 2015, p. 101 *et seq.*

42 On this subject, see Bönninger, "Wem gehören die Daten im Fahrzeug? Das moderne Fahrzeug – Messgerät, Steuergerät, Datenspeicher" in Deutscher Verkehrsgerichtstag (ed.), *52. Deutscher Verkehrsgerichtstag 2014*, p. 229 *et seq.*; Mielchen, "Verrat durch den eigenen PKW – wie kann man sich schützen?", *ibid.*, p. 241 *et seq.*; Rossnagel, "Grundlegende Rechtsverhältnisse und Ansprüche", *ibid.*, p. 257 *et seq.*; and Weichert, "Datenschutz im Auto", *ibid.*, p. 285 *et seq.*

43 Weichert, in Deutscher Verkehrsgerichtstag (ed.), *op.cit.* 2014, p. 291.

traditional distinction between personal data vs. non-personal data, taken together with the respective legal consequences, is robust enough to adequately protect the individual's right to informational self-determination.[44] A data protection and privacy law which in theory provides a very high level of protection such as that contained in the German Data Protection Act (BDSG) but in practice is frequently ignored[45] is insufficient if we consider the existing threats to data security. The State has a duty in practice, and not merely in theory, to effectively protect the right to informational self-determination, a right which remains exceptionally important. For this reason alone the entirety of German data protection law should be put on the test stand; space should be made next to traditional legal instruments for new models like "privacy by design" and "privacy by default".[46] In order to efficiently resolve problems of evidence arising in connection with automated driving, it would be sensible to have built-in data recorders become standard equipment for automated vehicles, as has been under discussion for decades.[47]

J. Insurance Law

Ultimately, the possible impact of the new technologies on the insurance industry could be quite significant.[48] One might predict that the accident rate will decrease, but that the severity of those accidents that do occur will increase significantly. At the same time, there will be a shift in liability from the human driver to the manufacturer of the automated driving

44 BVerfGE 65, 1 *et seq.* (Volkszählungsurteil = "public census case").

45 On the need to amend the law, see only Rossnagel, "Modernisierung des Datenschutzrechts" in Schmidt & Weichert (eds.), *Datenschutz. Grundlagen, Entwicklungen und Kontroversen*, 2012, p. 331 *et seq.*

46 On this subject, see for example Schaar, "Systemdatenschutz – Datenschutz durch Technik oder warum wir eine Datenschutztechnologie brauchen" in Schmidt & Weichert, *op.cit.* 2012, p. 363 *et seq.*

47 Schmidt-Cotta, "Event-Data-Recording – Fluch oder Segen?" in Hilgendorf, Hötitzsch & Lutz (eds.), *op.cit.* 2015, p. 67 *et seq.*

48 For an introduction, see Swiss Re Centre for Global Dialogue, *The autonomous car. Risks and opportunities for the re/insurance industry.* Conference report, September 2014.

system.[49] Insurance companies will have to take account of this shift when drafting insurance policies and calculating premiums.

What remains unclear is how the vulnerability of vehicle driving systems to sabotage will impact liability and insurance. If this sort of cybercrime is not countered with sufficient vigour by the law enforcement and criminal justice systems, we could face a situation in which the accident rate increases together with the severity of accidents, in contrast to the predicted decrease in the accident rate referred to above. In this context, the individual responsibility of the user, be that as a driver or as an owner of a vehicle, and its impact on liability have also been under discussion.

The issue of the usability of driving data by insurance companies is a familiar problem which is amplified, however, in the context of automated driving.[50] For an insurance company it could be quite interesting to know what distances a driver travelled, how fast the driver drove the vehicle, what style of driving the driver used as well as various physiological parameters such as respiration, heart rate, blood pressure, etc., which could be measured in the driver. In exchange for reductions in premiums, drivers could be given incentives to agree to the collection of such data. Such practices can, in principle, be economically efficient and often make sense from a health policy perspective as they can also encourage safer driving strategies. If conducted in an excessive manner, they would threaten the solidarity principle, which is the foundation upon which insurance is based. It might therefore be useful for the legislature to define a framework in which such practices could be conducted.

K. New Challenges

The overview presented above has dealt mainly with issues that are already acute today. Beyond that, there are a wide variety of problems

---

49 See above under F. For a comprehensive treatment of the topic, see Stadler, "Versicherungsrechtliche Fragen zu Fahrerassistenzsystemen" in Hilgendorf, Hötitzsch & Lutz (eds.), *op.cit.* 2015, p. 87 *et seq.*

50 The same problem has been under discussion in connection with so-called "wearables", i.e. devices worn on the body which, for example, collect and record physiological body data during sports activities. For more depth on this subject, cf. Bartmann, "Der kalkulierte Patient" in Schmidt & Weichert, *op.cit.* 2012, p. 178 *et seq.*

which, although not yet urgent today, should be addressed in the near future. The relevant keywords are:

I. Dealing with the "Control Dilemma"

The term "control dilemma" was used above to refer to the situation in which technological development allows the driver to be increasingly freed of monitoring and control tasks central to driving the vehicle, while the law still requires the driver to exercise almost complete control over the vehicle. This dilemma not only reduces the attractiveness of appropriately equipped cars (and trucks) significantly, but if left unsolved would also lead to overloading the driver, as psychological tests have shown that tasks demanding prolonged attention without stimulation cannot be performed. The question is of great practical importance, especially with regard to allowing drivers of autonomous vehicles to perform activities such as reading, writing, or using mobile phones, which would be dangerous distractions in normal driving.

In the current debate,[51] it is sometimes argued that a change in the law which would change the demands placed on the driver is not possible; this would violate the principle of culpability or Human Dignity or even the "Conception of Mankind" implicit in the German Federal Constitution.[52] In point of fact, it would be highly problematic to grant a statutory exemption to drivers of highly automated or fully automated vehicles, freeing them from all responsibility for the driving of the vehicle. In order to make the problem at issue manageable from a legal perspective, one needs to understand the following distinction: the demands placed by the law on the driver to exercise sufficient care are partly statutory and result partly from case law, i.e. the jurisprudence of the courts made by judges deciding individual cases.[53] The most important statutory source of law is the German Highway Code (StVO), but there are other statutory sources such as §§ 315c and 316 of the German Criminal Code (StGB).

---

51 Cf. above, pp. 183 *et seq*.

52 On this subject, see detailed treatment in Hilgendorf, *Konzeptionen des Menschenbilds und das Recht*, in Joerden, Hilgendorf & Thiele (eds.), *Menschenwürde und Medizin*, 2013, pp. 195 – 216.

53 Hilgendorf & Valerius, *op.cit.* 2013, § 12 paragraphs 18 *et seq*., and 20 *et seq*.

It is obvious that the legislature may amend previous statutes relevant to the issue of the duties of a driver on German roads. Nevertheless, it will be the task of the courts to decide what "sufficient care" is in each individual case and what kind of behaviour is considered as negligent. For German law, this rule reflects the constitutional principles of separation of powers (Art. 20 (2) 2 GG) and rule of law (Art. 20 (3) GG) rather than the principle of culpability, Human Dignity, or even the conception of mankind upon which our constitution is based.[54]

This rule has far-reaching consequences for automated driving: the legislature is free to amend statutory law on driving in order to adapt it to the new reality and possibilities offered by technology. For example, it could rescind the law prohibiting drivers from talking on mobile phones whilst driving (for drivers of vehicles equipped with sufficiently safe automated driving systems). It is obvious that this would need to be managed carefully so that it would take account of dangerous situations occurring in road traffic.[55] Additionally, it would be possible to amend the StVO to add a clause which stated that the duties of a driver on the road can be fulfilled not only by human drivers but also by means of appropriately powerful and secure technological systems.[56] The issues associated with case law may be among the biggest challenges automated driving holds for the law.

---

54 Under certain circumstances, the guarantee of a lawful judge, contained in Art. 101(1)(2) of the German Federal Constitution, could be relevant.

55 One could see it as a serious problem, for example, that for quite some time into the future, vehicles with very different levels of automation – including vehicles with only rudimentary automation – will be on the road. When confronted with a driver using his mobile phone whilst driving, the police have up to now been unable to determine what level of automation a vehicle has, just by looking at it from the outside.

56 A proposal was submitted by the Würzburg Research Centre on 26 February 2015 during the Ministry of Transport's roundtable on the legal problems of automated driving. According to the proposal, the following text would be included in § 1(3) of the new German StVO (highway code): "[1]Whoever carries out the task or a series of tasks of driving a motor vehicle, also complies with the requirements he has under this regulation, through the appropriate use of a vehicle system, equipped to carry out the driving tasks at issue. [2]The requirements of this regulation shall then apply in a corresponding manner to the vehicle driving system. [3]Whoever uses a vehicle system within the meaning of clause 1, may perform ancillary activities, if and to the extent that the activities are compatible in nature and extent with the intended use of the vehicle driving system.".

II. Writing Computer Algorithms that Prevent Accidents

A very difficult problem, which involves fundamental legal issues, arises in connection with accident algorithms, i.e. rules in computer programs designed to avoid damage and injury in accident situations. The starting point must be the prevention of losses and the minimisation of harm: if, for example, the vehicle can escape damage by avoiding the obstacle, this must be written into the program algorithm. If the vehicle is faced with the choice of minor property damage or major property damage, it must choose the former. The severity of the damage can also be determined using economic criteria.[57] But what is the system supposed to do if it has to choose between injuring two different persons? Should the decision be made based on the cost of medical care which would be incurred, or do other factors also play a role, such as long term care or consequential losses where the accident results in permanent disability[58]? How should the system decide if it has to choose between severely injuring several people or killing just one? The issue is to create a hierarchy of damage and injuries, which is feasible from a programming perspective and compatible both with our ethical notions and with the requirements of the law.[59]

This alludes to one of the classic textbook dilemmas in German law, in which several lives are pitted against each other: suddenly, two children appear in the lane of a high speed motorway, and next to them on the shoulder of the road is an adult. Unavoidably, either the children or the adult will be struck by the car and killed. How will / should / must the car's on-board computer decide? In German criminal and constitutional law, the opinion has prevailed since the 1950s,[60] that lives cannot be

---

57 This corresponds in principle with the procedures used in German law to calculate damages.

58 Of course, one could consider completely different criteria, such as age or gender. The absurdity of using such factors is quite obvious.

59 In Ethics, such problems have been under discussion for a long time within the context of Utilitarianism. That is outside the scope of this article. Allow me to note, however, that Bentham's theories would be more amenable to use in algorithms than Kant's (which, according to the views presented here, in no way speaks against Bentham).

60 For a comprehensive collection of citations, also covering the "switchman case", a classic textbook case in German law, as well as the "trolley cases", which are also discussed in common law jurisdictions, cf. Hilgendorf in Hilgendorf & Hötitzsch (eds.), *op.cit.* 2015, p. 20 *et seq*.

weighed against each other; one life is worth the same as 2, 20 or 200. Consequently it is the prevailing, and indeed almost unanimous opinion that even in an emergency (§ 34 StGB) it is illegal to kill one human being in order to save two others.[61]

But should this legal opinion hold sway, not in examining some event in hindsight, but in deciding on a course of action now in a real life situation, where lives can be saved? Is it really true that there is no difference between killing one person and killing several people? If so, then a Bavarian automobile manufacturer could program the accident algorithm of its autonomous vehicles so that in such emergency situations the car would kill the lowest number of pedestrians when the vehicle was travelling in Bavaria, but kill the highest number of pedestrians if the car should stray outside of Bavaria. It would be OK to pursue such a Pro-Bavaria Policy from an ethical and also from a legal perspective – if the issue of the number of lives – lost or saved – really was unimportant. The fact that this argumentation is strongly counter-intuitive (especially for non-Bavarians) is rather obvious. This problem cannot be discussed further here as it is outside the scope of this analysis. Random number generators don't really help, unless one is trying to avoid the problem of balancing lives against each other. Whether balancing lives against each other could ever be lawful seems doubtful. In my view, the killing of innocent human beings should remain unlawful, even in emergency situations. Nevertheless, there also exists a duty to minimise the wrong caused, which has the effect that there is not only an ethical but also a legal duty to minimise the number of persons killed.[62]

Such considerations are not merely scenarios to think about. Even if they seem to have more theoretical than practical significance, what is ultimately at issue here is the protection of a legal order where the highest values are acting with humanity and respecting Human Dignity.[63]

61 Hilgendorf &Valerius, *op.cit.* 2013, § 5 paragraph 83.

62 For further detail see Hilgendorf in Hilgendorf & Hötitzsch (eds.), *op.cit.* 2015, p. 26.

63 Hilgendorf in Horst Groschopp (ed.), *op.cit.* 2014, p. 39 *et seq.*

III. Further Problems in the Future

A number of future problems should be mentioned briefly here. One would be the possible legal duty to *switch software packages* when the automated vehicle crosses *national borders*. As long as the laws applicable to the driver and driving are different from country to country (and it is impossible to predict how long this will be the case), the on-board computer must be programmed so that it is able, upon crossing a border, to apply the relevant rules of the country where the car is located at that moment in time. An example of this would be the accident prevention algorithm. The software containing the driving rules would need to be exchanged at the point when the border was crossed. Perhaps such a software exchange could be performed automatically. The legal issues associated with this – going all the way from simple civil liability issues to questions of international law – have not even been clearly identified yet, let alone resolved.

Another problem would be that automated vehicles could collect data on the physiological well-being (i.e. *health data*) of their driver and possibly also of their passengers. This is already being done to a certain extent today by technological systems that warn the driver when he is tired. What should happen if the data which have been recorded reveal a long term deterioration in the health of the driver, such as a vulnerability to burnout? Must such data be stored (in principle, no)? Could some sort of guarantor relationship or duties arise, and for whom, if data collected in this way are also available for interpretation (which in any case cannot at the moment be excluded)? What are the consequences in terms of possible civil or criminal liability of the vehicle manufacturer[64]? What would also be controversial is whether the automatic transfer of such medical data into the Cloud would be lawful, just consider possible criminal sanctions under § 203(1) StGB.

Moreover, enormous challenges arise in the area of data protection and privacy when huge quantities of data are collected about a driver's driving behaviour, for use, for example, by a *predictive user experience* system. The creation of behavioural profiles based on the data collected is not some abstract danger inherent in data collection, but rather, and literally,

64 In principle, causing bodily harm by omission, here by the failure to get help in time, could possibly incur criminal liability under §§ 223, 13 StGB.

the intended purpose of the system.[65] How can this be reconciled with European and especially German data protection law, which among other things stress the principles of data minimisation and necessity? What rules apply if such collections of data are sold to third parties (data mining, "big data")? Is the data subject's consent necessary, and if so, how?[66]

Let me just note that predictive user experience systems, which provide the driver with tips about travel destinations of potential interest to him, also raise interesting ethical (and indirectly also criminal) questions. As the idea goes, while driving, the vehicle continually provides the driver with tips about sights of interest, shops, restaurants, and other locations. At first glance this seems totally unproblematic. But should the vehicle really point out the way to the chocolate factory to its obese driver without being solicited? Should it tell the alcoholic driver (and the machine knows this!) the way to the corner pub? It is possible that civil liability risks exist for manufacturers or distributors of these programs.[67] Should such programs perhaps at least adhere to the prevalent social morality of the place they are sold and, for example, refrain from providing tips about dubious sites (like, for example, betting parlours and brothels)? Should the vehicle be able to suggest short-cuts that violate the road traffic laws? Such questions allude to the topic of "moral machines", a theme which has already received some attention in the field of Ethics,[68] but which (so far!) holds hardly any relevance for legal practice. Therefore, let this brief reference suffice for now.

---

65 It is remarkable that such systems perform better and better, the more they know about their users. The inherent logic of unlimited data collection by definition renders meaningless the principles of data minimisation and necessity of data collection. Aggregating many sources of data then appears to be expedient (for example, individual shopping behaviour on- and offline, social networks, private emails, etc). Even the most intimate data on health issues could, under certain circumstances, help to improve system performance, making use of the system more attractive and more convenient.

66 Such case scenarios could be a reason to reconsider the limits of individual data sovereignty.

67 The judgments of some US courts in product liability litigation show that such a result is not as far fetched as one might think.

68 See for example Wallach (ed.), *Moral Machines: Teaching Robots Right from Wrong*, 2010, or Anderson & Anderson, *Machine Ethics*, 2011.

L. Summary

All in all, one could say that automated driving raises a large number of legal issues, which will keep lawyers and politicians intensely occupied in the years and decades to come. It is possible to infer a requirement to interpret existing law in a technology-friendly way from the German Constitution. The legislature is required, on the one hand, to amend the law in order to take account of new technologies, but on the other hand, also to expedite the introduction of new technologies which protect life and health through appropriate legislative measures.

A concrete need for amending legislation currently exists particularly in the area of road traffic law, by which domestic German law should, in the medium term, be adapted to the new provisions of the Vienna Convention. Amending the law on provider liability should, at the very least, be the subject of consultations. There is no need for legislation either in the area of civil or criminal liability. Having said that, the issue of the "control dilemma" raised by automated driving needs to be dealt with: technically it will very soon be possible for the driver to pursue activities other than the actual driving of the vehicle. Legally, however, the driver will probably remain under an obligation to monitor and control the vehicle for quite some time to come. Releasing the driver from monitoring and control activities takes us up to or potentially even beyond the limits currently set by the law, a problem which at the very least should soon be addressed. It is also very important to ensure adequate data and privacy protection. Finally, the harmonisation of law in the international context, which has not been discussed further here, is of great importance.

Delicate Decisions: Legally Compliant Emergency Algorithms for Autonomous Cars

Jochen Feldle,[1]
University of Würzburg

A. Introduction

No hands on the steering wheel, no foot on the gas pedal. Driving in the future will no longer be dependent on a human driver and could in this way become faster, safer and more eco-friendly. Automated driving is expected to be nothing less than the biggest revolution in transport since the invention of the automobile.[2] Almost every major car manufacturer has jumped onto the automation bandwagon. Google's driverless car, in particular, has attracted much public attention, as has the fatal accident of a Tesla Model S running on autopilot. By now it is a matter of common knowledge that not only technical, but also legal hurdles need to be cleared before fully automated vehicles (AVs) become a reality. Ethics is a contentious issue in this context. Although AVs could reduce traffic accidents – 90% of which are caused by driver errors[3] – even the most sophisticated models will crash[4] and someday inevitably be exposed to a "lose-lose situation". This could happen as follows: an autonomous car is travelling through an urban area, when suddenly a couple steps onto the street in front of it. Even if the vehicle makes an emergency stop, the pedestrians will be hit and most likely killed. A central barrier prevents the vehicle

---

1 Jochen Feldle is a Research Assistant at the *Forschungsstelle RobotRecht* (Research Center for Robotics Law), which is attached to the chair of Prof. Dr. Dr. Eric Hilgendorf at the University of Würzburg. His research focuses on the legal problems of autonomous systems.

2 Cf. the statement of Alexander Dobrindt, German Minister of Transport in *Die Zeit*, available at http://www.zeit.de/mobilitaet/2016-04/autonomes-fahren-gesetzentwurf-verkehrsrecht-alexander-dobrindt.

3 Statistisches Bundesamt (German Central Bureau of Statistics), *Verkehr auf einen Blick*, 2013, p. 38.

4 Noah J. Goodall, "Machine ethics and automated vehicles" in Meyer & Beiker (eds.), *Road Vehicle Automation*, pp. 93-102.

from swerving to the left. The car can only swerve to the right onto the sidewalk to avoid striking the couple. If it did that it would collide with an unsuspecting walker.[5] The car now has to choose between two evils, each of which involves killing people. Developing algorithms to deal with these kinds of emergencies is challenging, to say the least. In addition to legal issues, fundamental ethical questions also need to be addressed. This article focuses on just a few related problems.

B. Prevention of Harm

When a vehicle faces a desperate situation like this, why not have the vehicle hand control back to the driver and leave it to him? The human on board could serve as a "fallback" to take over driving whenever that was needed. For a number of reasons, this "solution" is rather unsatisfying. Our legal system[6] is designed to prevent all kinds of harm, if possible. Automated driving is supposed to make road traffic safer. Putting the passenger in charge in an irresolvable situation would be like a lottery. Simulation experiments suggest that, depending on the activity, human drivers need up to 40 seconds to gain awareness of the situation on the road.[7] A driver unaware of what is happening in his environment is not likely to make any decision at all. Of course, the benefits of automation would be nullified, if a hasty action by a human driver took precedence over a rational action by the onboard computer system. Furthermore, the absence of an appropriate decision algorithm could be seen as a latent construction flaw in the vehicle. The manufacturer could be held liable, both in civil

5 This kind of scenario has been discussed since the early 1950s, when Hans Welzel first devised the hypothetical case of a runaway train going down the railway tracks. A fatal collision with a passenger train – resulting in a large number of casualties – can only be prevented by pulling a lever, switching the train to a different set of tracks. But there are some railway employees working on the other track would be killed, Hans Welzel, *Zeitschrift für die gesamte Strafrechtswissenschaft* 63 (1951), p. 47 (51). British Philosopher Philippa Foot dealt with a very similar case in her article "The Problem of Abortion and the Doctrine of the Double Effect", *Oxford Review* 5 (1967), pp. 5 – 15. In English-speaking countries, the problem is known as the "Trolley Problem".

6 The analyses in this article refer exclusively to German law.

7 Patrick Lin, "Why ethics matters for autonomous cars" in Maurer, Gerdes *et al.* (eds.), *Autonomes Fahren*, p. 72.

law and in criminal law, for accidents attributed to such a defect.[8] If equipped with a powerful data processor, AVs can react to unexpected situations in a split second. In order for AVs to make the "right decision", "legal and ethical algorithms" will be required.

C. *Defining a Hierarchy of Legal Interests*

An AV constantly scans its surroundings with cameras, radar and lidar, amongst other things. Its movement follows the basic pattern "recognize – interpret – act".[9] For an operational system, each obstacle is just a "point on the radar", no matter whether it is a human, an animal or a lifeless object. In order to carry out their assigned tasks, AVs need an "idea" of how different legal interests are valued in our society. The underlying principle of § 34 German Criminal Code (*Strafgesetzbuch* – StGB) can serve as a starting point. According to the jurisprudence, a person, faced with imminent danger to life, limb or property,[10] who commits an act to avert danger to himself or another is justified if the affected legal interests *substantially outweigh* the ones interfered with. If a driver avoids a person, saving his life, but instead damages a parked bicycle, the elements of property damage are met. Yet his action is justified, so no criminal liability is incurred, since he has not injured the superior interest. This idea can serve as a basis for programming AVs.

Human life is commonly referred to as the highest value of our legal system.[11] That means that in the case of a collision, it prevails over every other interest.[12] Life is followed by physical integrity and then property.[13]

8 Eric Hilgendorf in Hilgendorf & Hötitzsch (eds.), *Das Recht vor den Herausforderungen der modernen Technik*, p. 22.

9 Frank Jourdan & Helmut Matschi, "Automatisiertes Fahren – Wie weit kann die Technik den Fahrer ersetzen?", *Neue Zeitschrift für Verkehrsrecht* 2015, pp. 26-29.

10 The section also mentions "freedom", "honor" and "other legal interest". Yet, it seems unlikely that an AV could harm these interests.

11 BVerfGE 39, 1 (42); BVerfGE 46, 160 (164); BGHSt 35, 347 (350); Fischer, *Strafgesetzbuch*, § 34 paragraph 14.

12 There are cases in which life could be subordinated to some other interest, e.g. in cases where the state could become susceptible to blackmail cf. Volker Erb in *Münchener Kommentar zum StGB*, § 34 paragraph 116. But those scenarios are hardly imaginable in the context of AVs.

13 Claus Roxin, *Strafrecht Allgemeiner Teil I*, § 16 paragraph 29.

In reality, the lines between fatal and (merely) serious injuries are blurred. Outside influences or simple coincidences can quickly change the fate of a potential victim. Naturally, every distinction only makes sense if machines are able to accurately predict the outcome of a collision.[14] Another issue that is still subject to discussion is whether minor injuries could be outweighed by large financial losses.

In addition to protecting the interests mentioned above, AVs need to follow the applicable traffic laws and rules of the road. These sets of rules could either be treated as "hard constraints" or "soft constraints".[15] The consequences for vehicle decision-making can easily be illustrated. While *hard constraints* would function as impenetrable "virtual walls", *soft constraints* could be integrated into a hierarchy of interests. The disadvantages of absolute restrictions are quite obvious. Imagine an imminent collision of an AV with a pedestrian at high speed. If crossing a solid line is as equally prohibited as swerving onto a sidewalk, the car cannot avoid the collision – even if there is no road user or other pedestrian for miles around. Such a program would create unnecessary lose-lose situations. Furthermore, in normal traffic flow, AVs might block traffic, since they could not, for example, pass an abandoned vehicle parked on the street, if in doing so they had to cross a solid line.

But how should traffic laws and rules of the road be ranked? When weighing different interests within the framework of § 34 StGB, administrative concerns are less important than concrete legally protected interests.[16] Traffic rules could be suspended if their application would lead to unbearable consequences.[17] Traffic regulations should make transport safer. Naturally, they cannot be placed above human life or the integrity of the human body. It seems appropriate also to tolerate breaches of rules to save property. Even where this involves less valuable property, this seems well justifiable. If avoiding a collision (which always carries some danger) involves no risks for other traffic participants, one should not insist on compliance with traffic rules at the expense of another legally protected

14 Criteria could be speed, strike zone and statistical values, as well as safety systems such as pedestrian airbags.

15 Christian Gerdes & Sarah Thornton, "Implementable Ethics for Autonomous Vehicles" in Maurer, Gerdes *et al.* (eds.), *Autonomes Fahren*, p. 98 *et seq*.

16 Volker Erb in *Münchener Kommentar zum StGB*, § 34 paragraph 112; Ulfrid Neumann in *Nomos-Kommentar zum StGB*, § 34 paragraph 5 *et seq*.

17 Ulfrid Neumann in *Nomos-Kommentar zum StGB*, § 34 paragraph 5 *et seq*.

interest. As long as the rules of the road cannot be outweighed by personal concerns, such as time pressures, a general "brutalization of traffic manners" is not to be expected.

D. Collision of two Legal Interests of Equal Weight

When two interests on different levels collide, the legislature has provided a simple rule to follow: the interest *substantially outweighing* the others comes out on top. Strictly speaking, these situations present pseudo-dilemmas,[18] since our legal system has a solution at hand. But if two interests of the same level collide, knowing their abstract place in the hierarchy is insufficient. AVs could consider the decision unimportant and act randomly. Even if only chattels are at stake, this could be problematic. A football is a chattel, as is a parked car or some priceless and irreplaceable painting. Of course it is evident that they should be treated differently. When it comes to chattels, it is totally reasonable for the decision to be made based on value or quantity. The issue here is minimizing damage. Generally, this also applies to injuries; the risk of severe injuries should outweigh the risk of only slight injuries.[19] But when it comes to the weighing of lives against each other, it is not that simple. Recall the scenario outlined at the beginning of this paper, which is often reduced to a simple conflict between utilitarian and deontological ethics. The former counts lives, the latter does not. There are various scenarios and almost as many ways to solve them. The car could go straight and run over two pedestrians or swerve to the right killing one. It could also crash into the central barrier sacrificing itself and its passengers.[20]

I. Decision by Numbers

The absolute numbers could be the crucial factor. Choosing the *smaller number* is sometimes used synonymously with choosing the *lesser evil.* The American legal system does it this way. If a person who is confronted with an irresolvable situation commits what would normally be considered

---

18 Cf. Harro Otto, *Juristische Ausbildung* 2005, p. 470 (471).

19 If accurate predictions are possible.

20 This delicate matter is not within the scope of the present paper.

a crime, he may be justified by necessity. The defense of necessity is available if a person intentionally kills another individual in order to save two or more people.[21] However, the German approach differs fundamentally from this. According to German court jurisprudence on the German Constitution, every life represents a “maximum value”. Weighing lives against each other is therefore strictly prohibited. The Federal Supreme Court *(Bundesgerichtshof)* clarified this many years ago and has never deviated from this principle.[22] More recently, the Federal Constitutional Court *(Bundesverfassungsgericht)* held that the sacrifice of lives in order to save other lives violates Human Dignity, the first and most important principle in our Constitution.[23] Therefore, in 2006 the Court declared that § 14 (3) of the Aviation Security Act *(Luftsicherheitsgesetz)* was unconstitutional. This act was passed as a reaction to the 9/11 attacks and would have authorized the German Federal Armed Forces to shoot down a hijacked passenger airplane if it was being used as a weapon to kill others. As § 34 StGB shows, every person must accept some interference with his legal interests. It can also be expected that people will act with solidarity toward one another. But a line is drawn between what can be tolerated and what cannot be tolerated when a person needs to sacrifice his life in order to save another’s life.

Despite this settled case-law, some consider absolute numbers of people to be a crucial factor.[24] Imagine that a hospital in a secluded provincial town is on fire. The only fire truck from the local fire department can either extinguish the fire in the east wing, with 5 persons in it, or in the west wing, with 20 persons in it.[25] If two equal duties are in conflict, the person under the duty normally has the right to choose and is justified either way. In this particular case, however, this right could be restricted. It is the view of some jurists that an action is less negative where it saves many lives.[26] It would follow logically that it is better to save more lives than less so one could well assume that not equal but rather unequal duties collide.

21 Wayne LaFave, *Criminal Law*, p. 557.

22 BGH, *Neue Juristische Wochenschrift* 1953, p. 513 (514).

23 BVerfGE p. 115 (118 *et seq*). .

24 Cf. Franz-Benno Delonge, *Die Interessenabwägung nach § 34 StGB und ihr Verhältnis zu den übrigen strafrechtlichen Rechtfertigungsgründen*, p. 118 *et seq*.

25 This fictional case was presented by Merkel, *Juristen Zeitung* 2007, p. 373 (380).

26 Wolfgang Mitsch, „Nantucket Sleighride“ – Der Tod des Matrosen Owen Coffin, in *Festschrift für Weber*, p. 49 (64).

Saving the larger number of persons would therefore be preferable. Some commentators consider this solution to be right and proper.[27] If this is correct, it follows that where the person under a duty saves the lesser number of persons, this could not be justified. This opinion is held by a minority. The victims in the east wing become an abstract number rather than human beings. According to the Federal Constitutional Court that is precisely what violates Human Dignity. If the fundamental values of our Constitution are taken seriously, there can be no distinction based on numbers. Naturally, the person under a duty can base his decision on the numbers. Not because one duty outweighs the other, but because the duties are equal so that either choice is correct. A decision like this is not illegal. Legally speaking, saving the larger number of persons cannot be imperative. Depending on one's point of view, moral responsibility might also play a role, but saving fewer people instead of more people does not make the justification void.

Therefore, autonomous cars may not swerve to one side, killing people, in order to save other people. In order to violate the legal interests of those being killed, one would need a higher interest. But *life* represents the highest interest in the hierarchy. Two lives cannot outweigh one and neither can a thousand outweigh a hundred. Every single life represents a "maximum value". In other words, the value of one life is infinite. Regardless of the number of lives at stake, the value will not increase ($100 \times \infty = \infty$). Questioning the infinite value of a single life violates Human Dignity. The AV in the example discussed at the beginning of this paper must go straight and kill the couple. Swerving to the right cannot be justified.

II. Decision via a Random Generator

Every alternative concept based on numbers risks being attacked as unconstitutional. Therefore, random number generators have been suggested in order to resolve the problem.[28] In place of complex – and disputable – ethical algorithms, chance could decide who gets to live. Fortune is not alien to our legal system. If there are too many applicants for places at uni-

---

27 Reinhard Merkel, *Juristen Zeitung* 2007, p. 373 (380).

28 This idea is picked up by Volker Johanning/Roman Mildner, *Car IT kompakt*, p. 73; Patrick Lin, "Why ethics matters for autonomous cars" in Maurer, Gerdes *et al.* (eds.), *Autonomes Fahren*, p. 71.

versity, the Federal Constitutional Court has acknowledged that a "decision by lots" is lawful.[29] But randomized decisions about life and death are frequently considered to be offensive. Some have compared this kind of decision making to a "life-lottery".[30] However, in cases of other life-and-death situations, such as organ transplantation, a randomized pick has been considered a promising approach.[31] According to § 12 (3) (1) German Transplantation Law (*Transplantationsgesetz* – TPG), the organs subject to allocation are distributed according to chances of success and urgency. In order to ensure true equality, organs could be distributed randomly within a group of patients (e.g. high-risk patients). Indeed, this method need not be seen as some kind of "divine judgment",[32] but could be considered the most egalitarian system of allocation. Chance does not discriminate against anyone. Everyone has the same opportunity to benefit from getting the transplant.

In today's traffic situations, a human driver is not able to perceive all the factors relevant to making a decision. His actions could be said to be more or less random and most likely do not incur criminal liability. Modern society has to accept this. So why not program AVs to behave in exactly the same manner? Everyday road traffic would still be safer. Emergencies, which occur rarely, would be treated as they always were. All things considered, this would be a success for autonomous driving, wouldn't it?!

Randomized decisions seem reasonable, if all "rational" criteria fail. A comparison can be made to a person lost in the woods.[33] He needs to pick a direction and follow it through if he does not want to starve. He simply cannot know whether he has chosen the "right" path. In contrast, programmers who write code for AVs are not "lost". Algorithms need to be in line with statutes and Supreme Court jurisprudence, which are based on the values of our Constitution. Furthermore, in the case of organ transplantation and other situations where randomized decisions are discussed (such as triage in emergency situations), the chances of survival are distributed equally. For this reason, a decision by lots can bring true equality of op-

---

29 BVerfGE, *Neue Juristische Wochenschrift* 1977, p. 569 (572).

30 In the context of organ transplantation, cf. Günter Feuerstein, *Das Transplantationssystem*, p. 253.

31 E.g. Mathis Bader, *Organmangel und Organverteilung*, p. 507 *et seq*. (with limitations).

32 Klaus Bernsmann, *Entschuldigung durch Notstand*, p. 336.

33 Otto Depenheuer, *Juristen Zeitung* 1993, p. 171 (176).

portunity. But the AV's decision would endanger a pedestrian who had previously been safe. When a vehicle departs from its original trajectory, the risks are reallocated. This contradicts the idea of equal opportunity. Furthermore, the comparison to a human driver is poor. Legal issues arise from the fact that humans operate machines the speed of which can easily exceed their ability to react. Things happen too fast for the human body to keep up, whereas AVs equipped with powerful data processors can. Replacing a lawful solution with a random generator would obscure who is meant to be responsible in order to avoid an admittedly difficult discussion.

E. Conclusions

Important decisions in daily life are already being made using algorithms. In Germany, so called credit scoring is widely accepted. Based on general information such as "profession" or "place of residence", a numerical expression is defined, representing the creditworthiness of a person.[34] Naturally, it feels strange when machines – such as AVs – have the power to decide between life and death. This fear has been used to scandalize the issue far before the rollout of the first generation of AVs with such decision-making algorithms. *Der Spiegel* featured the headline "The lottery of death",[35] failing to recognize how these algorithms could work and which functionalities they will deliver. The Federal Constitutional Court holds a very firm position on the weighing of lives. Car makers cannot ignore this jurisprudence. For this reason, random generators will remain an idea that cannot be implemented. Emergency algorithms for autonomous cars must be consistent with the fundamental values of our legal system, which state: every life has equal value. Lives cannot be balanced against each other. One cannot simply bypass making a decision by using random generators. The actions of AVs will therefore need to be predictable. This will not be a lottery, but the application of established law.

34 Based on this score, such things as interest rates are determined.

35 Manfred Dworschak, "Die Lotterie des Sterbens", *Der Spiegel*, 04/2016, p. 104 *et seq.*

Flying Between the Lines: Drone Laws and the (Re)Production of Public Spaces

Kristen Thomasen,[1]
University of Windsor (Ontario)

"In enacting law, we enact space, and vice versa."[2]

A. Introduction

Remotely-piloted unmanned aircraft systems, colloquially known as "drones",[3] have existed since the early 1900s.[4] Only recently though have advances in aircraft and sensor technology made drones less expensive to

---

1 Assistant Professor of Law, Robotics and Society at the University of Windsor Faculty of Law, Doctoral Candidate in Law at the University of Ottawa, Canada. Thank you to Eric Hilgendorf, Uwe Seidel and Ute Rosin for inviting me to participate in the Automatica International Law Workshop; to my PhD supervisor Ian Kerr for his constant support and encouragement; and to Katie Szilagyi and John Popham for their helpful insights and comments. And thank you to the Social Sciences and Humanities Research Council for funding this research.

2 Nicholas Blomley, "From 'What' to 'So What': Law and Geography in Retrospect" in Jane Holder and Carolyn Harrison (eds) *Law and Geography* (Oxford University Press, 2003) 17-34 at 31 [Blomley, "So What"].

3 The technology generically referred to as a "drone" or "aerial drone" can also be referred to more specifically by various technical terms. For the sake of both a broad and inclusive denotation, as well as ease of readability, I have used the term "drone" throughout this paper to refer to the variety of unmanned aerial vehicle systems that exist. For further reading on terminology and the nuanced meanings of various terms, see for example, Roger Clarke, "Understanding the Drone Epidemic" (2014) 30 Computer Law & Security Review at 230-246. The Australian Certified UAV Operators Inc. also provides a basic description of drone terminology here: http://www.acuo.org.au/industry-information/terminology/what-do-we-call-them/.

4 Jonathan B. Rupprecht and Donna Dulo, "A Brief History of Unmanned Aircraft" in Donna Dulo (ed) *Unmanned Aircraft in the National Airspace* (American Bar Association, 2015) pp. 29-39 at 32.

operate and easier to fly.[5] This has increased their accessibility to state, commercial and private actors; a trend that is expected to continue.[6] The already growing drone market is projected to expand to an over $100 billion dollar commercial industry worldwide in the next five years.[7]

With this surge in popularity, drone technology will increasingly operate at a challenging nexus of law and space, as the laws that regulate drone technology and domestic airspace are amended or interpreted to permit wider drone use within that space.[8] Some of the most exciting and commercially promising applications of drones – from film, photography, and aerial observation, to the delivery of goods and people – will take place in low-altitude airspace.[9] This is by design. The laws regulating the use of commercially accessible drones logically require that the technology be operated both safely below commercial airline operations, and safely above (or away from) people and buildings.[10] By extension, these regula-

5 See e.g. Donna Dulo, "Aeronautical Foundations of the Unmanned Aircraft" in Dulo, *Unmanned Aircraft* at pp. 41-56; Roger Clarke, "Understanding the Drone Epidemic" (2014) 30 Computer Law & Security Review 230-246.

6 Scott Thompson, and Alana Saulnier, "The "Rise" of Unmanned Aerial Vehicles (UAVs) in Canada: An Analysis of Special Flight Operation Certificates (SFOCs) from 2007 to 2012" (2015) 41 Canadian Public Policy 207-222.

7 Wojciech Moskwa, "World Drone Market Seen Nearing $127 Billion in 2020, PwC Says" May 9, 2016, Bloomberg Technology: http://www.bloomberg.com/news/articles/2016-05-09/world-drone-market-seen-nearing-127-billion-in-2020-pwc-says.

8 In regards to this nexus of law and space more generally, see e.g. Nicholas K. Blomley and Gordon L. Clark "Law, Theory, and Geography" (1990) 11 Urban Geography 433–446 at 435; Luke Bennett and Antonia Layard, "Legal Geography: Becoming Spatial Detectives" (2015) 9 Geography Compass 406-422 [Bennett and Layard].

9 For the purposes of this paper, low-altitude airspace refers to the airspace above the ground and below approximately 500 feet, above which airspace is designated for manned aviation. See e.g. Troy Rule, "Airspace in an Age of Drones" (2015) 95 Boston University Law Review 155-208 [Rule, "Airspace"], discussing the U.S. approach to regulating airspace, mandating drones remain below manned aircraft.

10 Section II examines some specific statutory examples. It is also possible that domestic drones will operate above commercial airspace (see for example the Facebook Connectivity Lab: "Connectivity Lab", Internet.org, accessed January 2017: https://info.internet.org/en/story/connectivity-lab/), however this is less technologically imminent and not the focus of this paper. This paper also does not address the military use of drones, which falls within a different legal framework and raises additional legal, ethical and spatial issues.

tory boundaries place many current and emerging drone operations within a space that, to date, has not been heavily utilized by people or by other technologies.[11] The growing domestic prevalence of drones and their increasingly diverse range of applications have prompted state regulators and lawmakers, academics, and drone operators alike to attempt to better conceptualize this newly accessible aerial space. But the way in which this airspace is understood, particularly through the laws that regulate drone-based access to low-altitude airspace, could have significant, potentially unforeseen and sometimes competing consequences that go beyond efforts to maintain a safe, secure and efficient airspace.[12]

The field of Law and Geography provides an insightful lens through which to identify some of the less immediately obvious, but socially significant, impacts of drone regulations. Legal geography is concerned with the mutually constitutive relationship between law and space. In particular, legal geographers illuminate the ways in which laws can inform, create or shape lived, real world spaces and our experiences within those spaces.[13] These insights encourage more explicit consideration of the ways in which drone laws – those determining who can fly a drone, where and for what purpose – can have the effect of producing new (air)spaces, and re-shaping existing (ground) spaces, as well as regulating access to, control over, and relative power within these spaces. Examining the ways in which these laws might produce newly accessible, potentially public spaces encourages a more holistic consideration of the values that should inform national and local drone regulation, from both the perspective of drone operators and the individuals who encounter drones.

---

11 While other aerial technologies have access to these spaces (for example, a plane or helicopter can technically hover at lower altitudes, including over private property, see e.g. *Dow v United States*, 226 F. 145 (4th Cir., 1915) [*Dow*]; *Florida v Riley*, 488 US 445 (1989) [*Riley*]; *California v Ciraolo*, 476 US 207 (1986) [*Ciraolo*]) the low cost and unmanned characteristics of drones permit a wholly different use of this space. For a deeper discussion of the features that set drones apart from other aerial technologies in this context, see Section III below.

12 See e.g. Transport Canada, "What We Do", accessed January 2017: https://www.tc.gc.ca/eng/aboutus-whatwedo.htm; Federal Aviation Administration, "Mission and Responsibilities", accessed January 2017: https://www.faa.gov/airports/central/about_airports/CE_mission/.

13 David Delaney, "Beyond the Word: Law as a Thing of this World" in Jane Holder and Carolyn Harrison, Law and Geography (Oxford: Oxford University Press, 2003) pp. 67-83.

The central focus of this paper is, simply, to consider some initial insights about drone regulation that can be drawn from the field of Law and Geography. Ultimately the paper aims to recognize this field of scholarship as a source of deeper critical interrogation for those concerned with the development of domestic drone rules and regulations. In the sections that follow, I first set out some theories and insights from Law and Geography that can help to illuminate some important social consequences of drone regulations. I subsequently provide a more detailed explanation of the ways in which airspace laws have already produced a public airspace, the boundaries of which are being challenged through the integration of drones into low-altitude airspace. The paper concludes with examples of the potential application of legal geography's analytical tools to drone laws, identifying some ways in which drone laws can produce, or reshape, the experience of public spaces.[14]

B. Law and Geography

Law and Geography is an interdisciplinary field of critical legal scholarship dedicated to identifying and untangling the co-constitutive relationships between law and space.[15] Essentially, legal geographers contend that space does not simply exist; it is *produced*, through legal mechanisms.[16] Space, simultaneously, shapes the law, for example, in the ways that juris-

14 The fact that drones render a low-altitude airspace newly accessible to the public, and therefore catalyze legal treatment of that space, is only one of many potential spatial dimensions of drone technology about which Law and Geography might provide helpful future insights. See for example, Ryan Calo, "Robotics and the Lessons of Cyberlaw" (2015) 103 California Law Review 513-563 at 532-537, on the legal significance of the physical nature of robotic technologies generally.

15 Antonia Layard, "Freedom of Expression and Spatial (Imaginations of) Justice" in Dimitry Kochenov, Grianne de Burca, Andrew Williams (eds), *Europe's Justice Deficit?* (Oxford: Hart Publishing, 2015); Bennett and Layard, "Spatial Detectives" at 406.

16 Building from Henri Lefebvre's, *The Production of Space*,Donald Nicholson-Smith (trans.) (New Jersey: Wiley-Blackwell, 1992), see e.g. Chris Butler "Critical Legal Studies and the Politics of Space" (2009) 18 Social & Legal Studies 313-332; Nicholas Blomley, "Law, Property and the Spaces of Violence: the Frontier, the Survey, and the Grid" (2003) 93 Annals, Association of American Geographers 121–141; Iris Braverman, "Who's Afraid of Methodology? Advocating a Methodological Turn in Legal Geography" In Braverman, I., Blomley, N., De-

diction and federalism can determine the lawful authority to act.[17] Further, space and law are not neutral concepts, but rather both are shaped by and serve to shape society, social relations and experiences, and dynamics of power.[18] Ultimately, legal geographers "find that law is (and has always been) embedded in the world rather than hovering above it in some disembodied way."[19]

Legal geography provides tools through which we can identify and assess some of the significant and complex relationships between law, space and domestic drone technology, which might otherwise go unseen.[20] One such concept is referred to as the "splice." Coined by Professor Nicholas Blomley, a "splice" is an occurrence that simultaneously embodies legal and spatial meaning.[21] An example of a splice, which is particularly relevant in the context of domestic drones, is the concept of 'public space.'[22] This splice simultaneously embodies legal, spatial and social signifi-

---

laney, D. and Kedar, A. (eds) *The Expanding Spaces of Law: a Timely Legal Geography* (Stanford University Press: Stanford, 2014); Bennett and Layard at 408.

17 Nicholas Blomley and Joel Bakan, "Spacing Out: Towards a Critical Geography of Law" (1992) 30 Osgoode Hall Law Journal 661-690.

18 Layard, "Freedom of Expression", *supra* at 4, citing C. Schmid "Henri Lefebvre's Theory of the Production of Space: Towards a Three Dimensional Dialectic" in K Goonewardena et al (eds) *Space, Difference, Everyday Life: Reading Henri Lefebvre* (New York: Routledge, 2008) at 28. See also S. Blandy and D. Sibley "Law, boundaries and the production of space" (2010) 19 Social & Legal Studies 275–284; Bennett and Layard at 409.

19 Bennett and Layard *ibid* at 417; Delaney, "Beyond the Word", *supra*; Layard, "Freedom of Expression" *supra*; Blomley, "Law, Property and the Spaces of Violence", *supra*.

20 While this chapter focuses on the domestic use and regulation of drone technology, Law and Geography scholarship has also interrogated legal/spatial dimensions of military drone use. See e.g. Campbell A.O. Munro, "Mapping the Vertical Battlespace: Towards a Legal Cartography of Aerial Sovereignty" (2014) 2 London Review of International Law 233-261; Kenneth Anderson, "Targeted Killing and Drone Warfare: How we Came to Debate Whether there is a "Legal Geography of War"" in Peter Berkowitz (ed), *Future Challenges in National Security and Law* (Hoover Institution Press, 2011).

21 Bennett and Layard "Spatial Detectives", *supra* at 410. See also Delaney's concept of the "nomosphere" in D. Delaney, *The Spatial, the Legal and the Pragmatics of Place-Making: Nomospheric Investigations* (Routledge: Abingdon, 2010).

22 The meaning of "public" in this context can be understood through reference to the legal status, public access, values, and uses that render a space public. For more see e.g. Lefebvre, *The Production of Space*; S. Ruddick, "Constructing Difference in Public Spaces" (1996) 17 Urban Geography 132–151; Don Mitchell,

cance.[23] And the legal, spatial and social understandings of the space serve to co-construct the space. For instance, as Professor Antonia Layard explains, the physical qualities of a space, such as being accessible by members of society, do not alone determine whether a space is a 'public' space. A shopping centre may feel public to those seeking to enter, but it is legally designated as private property and accordingly its owners have a property-based right to exclude.[24] Accordingly, to actually be a *public* space splice, the space also requires the legal designation of being 'public', and the protections that flow from that designation.[25] But also, simply designating a space as 'public' in law does not, on its own, render a space truly public. For the space to *be* public, members of the public must be able to identify it as such, and must be able to access and use the space.[26] Accordingly, a public space splice emerges from a co-constituting relationship between the spatial characteristics of the site, which permit access, and its legal designation as public, which protects the rights of those who enter to be in that space.

Furthermore, beyond the legal designation of a space as public, the regulatory regimes that apply in that public space may also determine the true public nature of the space.[27] A regulatory regime that excludes certain individuals, conducts, or activities can render a public space *less* public, to the extent that the space can no longer be communally accessed, or used by all members of the public.[28] This may also mean that different individuals can experience the same space as either public or private.[29] Accordingly, in addition to considering the private or public ownership status of a space, legal geography calls for a deeper consideration of the regimes that regulate the public nature of a space, in order to assess the extent to which it is in fact a "public" splice.[30]

The Right to the City; Evelyn Ruppert, "Rights to Public Space: Regulatory Reconfigurations of Liberty" (2006) 27 Urban Geography 271-292.

23 See e.g. Mitchell, *The Right to the City, supra*.

24 See Layard, "Freedom of Expression", *supra* at 5; see also *Harrison v Carswell* [1976] 2 SCR 200.

25 Layard, *ibid* at 6.

26 Layard, *ibid* at 7.

27 Mitchell, *The Right to the City, supra* at 35.

28 Ruppert, "Rights to Public Space" *supra*.

29 Blomley, "So What", *supra* at 23.

30 See e.g. Ruppert. "Rights to Public Space", *supra*.

Another tool within legal geography that can facilitate further analysis of drone law is the concept of a "legal-spatial imaginary." This concept refers to the manner in which a judge or lawmaker understands, or imagines, a space.[31] It is again premised upon the fundamental notion that space is constructed, emphasizing that there are multiple ways of seeing and understanding a given space.[32] The legal decision-maker's understanding of how a particular space should look, be, or feel, has powerful effects when translated into law, shaping the experience of that space with legal force.[33] One of the goals in legal geography is to identify and unpack these legal-spatial imaginaries, in order to understand their effect in producing social spaces.[34] Drawing from even just these two concepts, we can start to identify some of the spatial and social implications of domestic drone laws. The following sections of this paper seek to do exactly this – beginning with an exploration of some of the ways in which public space has already been produced and understood through the development and application of national airspace laws.

C. The Production of a "Public" Airspace

Drawing upon the theoretical tools set out above, the following sub-sections explore some of the ways in which technology-mediated access to airspace has correlated with the adoption of laws that purport to produce a public airspace. However, as noted above, a space is not truly a public space "splice" if it lacks public access or a legal designation as public, and may furthermore be rendered less public through laws that differentially regulate who may access that space, and which activities may take place there.

The first sub-section below considers how public airspace came to be a "public" space – through the co-constituting interaction between a newly accessible airspace and legal decision-making. Drones now enter into this public airspace. But paradoxically, increased public access to the airspace,

---

31 Layard, "Freedom of Expression", *supra* at 8.

32 J Allen, D Massey and A Cochrane, *Rethinking the Region* (London, Routledge, 1998) 1-2.

33 Layard, "Freedom of Expression", *supra* at 8.

34 Layard *ibid* at 5; Bennett and Layard, "Spatial Detectives", *supra* at 412; Delaney *Place-Making*, *supra.*

made possible through drone technology, is causing the legal designation of airspace to be called back into question. Furthermore, the regulation of drone activities within that space challenges the notion that it is in fact "public". These issues are the focus of the second and third sub-sections below. The final sub-section considers one potential impact that a public air "splice" may have on public space on the ground, with the possible effect of rendering it less public to some who utilize that ground space. Ultimately, these examples serve to highlight aspects of drone laws and uses that require further consideration by the legal decision-makers who will conceptualize and regulate this space.

I. From Private to Public Airspace

Drone technology is, of course, not the first technology to enter into airspace.[35] This section explains how advances in technology have already catalyzed a legal conceptualization of the airspace, laying the groundwork for the integration of drones. While it is beyond the scope of this paper to consider worldwide approaches to airspace regulation, this section considers demonstrative legal examples from Canada and the United States.

Prior to the advent of manned aviation, common law in Canada and the U.S. conceived of the airspace above private land as an extension of that

35 Airspace around the world is heavily regulated to facilitate safe and coordinated access to and use of this space by manned aircraft. Civil airspace is regulated internationally through for example the *Convention on International Civil Aviation,* 7 December 1944 (entered into force 4 April 1947) (also known as the *Chicago Convention*). The *Chicago Convention* sets out the basic principles for international civil aviation, including first and foremost that "every State has complete and exclusive sovereignty over the airspace above its territory" (Article 1). Further regulation occurs on the multilateral and national levels. This is in contrast to outer space, which is internationally regulated by treaty, see in particular the *Treaty on Principles Governing the Activities of States in the Exploration and Use of Outer Space, including the Moon and Other Celestial Bodies (Outer Space Treaty)* (December 19, 1966). While drones may become capable, technologically and potentially legally, of operating in outer space, that is not the focus of this paper because drones at such high altitudes do not (at least yet) have the immediate impact on the lived experiences on and from the ground, as do drones at lower altitudes. For more on potential high-altitude commercial drones, see for example the Facebook Connectivity Lab: "Connectivity Lab", Internet.org, accessed January 2017: https://info.internet.org/en/story/connectivity-lab/.

property; legal doctrine presumed that possession rights in the land extended to the reaches of the universe.[36] In other words, the airspace above a property was legally treated as an extension of the private property below. The landowner accordingly had the power to permit the entry or exclusion of persons and things within this private space.[37] Of course, at this time airspace was not actually widely accessible by members of the public.

However, with the arrival of manned aviation, this legal designation of airspace risked imposing a significant legal barrier to the development and implementation of a socially and economically transformative industry.[38] Under the historical legal doctrine, airlines might have to seek permission from each landowner whose property their planes might pass over – the logistics of which, let alone the cost, would be prohibitive. Accordingly, this new technological reality, in which broader use of airspace had become possible, prompted a now widely affirmed change in the legal conceptualization ("imaginary") of airspace, doing away with the former notion that the airspace up 'to the heavens' could be subject to private ownership.[39]

For instance, in the United States, Congress specifically designated "navigable airspace" – airspace more than 500 feet above the ground – as falling within the public domain.[40] The United States Supreme Court affirmed this legal designation in *United States v Causby*.[41] The Court held that navigable airspace was to serve as a "public highway" for aviation.[42]

36 In accordance with ancient legal maxim: *Cujus est solum, ejus est usque ad coelom* ("He who owns the soil owns everything above [and below] from heaven [to hell]"). See e.g. Troy Rule "Airspace" *supra*, for a more detailed background to this doctrine.

37 Stuart Banner, *Who Owns the Sky? The Struggle to Control Airspace from the Wright Brothers On* (Cambridge: Harvard University Press, 2008) at 16-19.

38 See e.g. Banner, *Who Owns the Sky?, supra.*

39 See e.g. Banner, *Who Owns the Sky?, supra*; Kevin Gray, "Property in Thin Air" (1991) 50 Cambridge Law Journal 252-307; *United States v Causby*, 328 US 256, 260-61 (1946) [*Causby*].

40 *Air Commerce Act of 1926*, Pub. L. No. 69-254, 44 Stat. 568, and later amended in *Civil Aeronautics Act of 1938*, ch. 601, 52 Stat. 973.

41 *Causby*, 260-61.

42 *Ibid.*

Landowners could not exclude aircraft from that space.[43] Douglas J., writing for the court, went so far as to emphasize that "[c]ommon sense revolts at the idea" that the airspace would be "clog[ged]" by private claims to ownership, underscoring his conceptualization of airspace as a space for efficient 'public' transit.[44]

Canadian courts came to a similar determination. For example, the court in *Didow v Alberta Power* adopted the approach set out in *Causby*, and a similarly reasoned leading British case, *Bernstein v Skyviews*, concluding that a landowner "has no greater rights in the air space than any other member of the *public*."[45] The court further emphasized that the public nature of airspace was in the public interest, to "take advantage of all that science now offers in the use of air space."[46]

These are but a few leading examples highlighting the ultimate observation that, as airspace became more physically accessible to the public,[47] the legal conceptualization and designation of that space changed, facilitating further public access to that space. With these above decisions, the 'splice' of airspace accessible by manned aircraft changed from being generally private, to a largely public space. Airspace laws produced a new public space.

II. The Tenuous Legal Designation of Low-Altitude Drone-Accessible Airspace

Despite the above, while Canadian and U.S. airspace has now long borne a legal designation as "public," genuine wide-spread public access to that space has for most of that time been practically limited to commercial aviation. In particular, aircraft are expensive to own and complicated to oper-

43 See also the discussion of this case in A. Michael Froomkin and P. Zak Colangelo, "Self-Defense Against Robots and Drones" (2015) 48 Connecticut Law Review 1-69 at 24.

44 *Causby* at 261.

45 From *Bernstein of Leigh (Baron) v Skyviews & General Ltd.* [1978] Q.B. 479 at 488B, followed in *Didow v Alberta Power Ltd,* [1988] 5 WWR 606 (ABCA) (italics added).

46 *Bernstein* at 907 quoted in *Didow* at para. 36.

47 This was and still is a narrow definition of 'public' – including only those who can, or can afford to, access this space through aircraft. Drone technology, by contrast, stands to be much more widely financially accessible.

ate, typically requiring significant training and infrastructure. Relatively few individuals own personal aircraft. Even commercial aviation can be relatively expensive to access. Low-altitude airspace has been even less accessible. The safe operation of manned aircraft often requires that flights remain at high altitudes, with the exception of takeoff and landing.[48] In fact, higher-altitude flights are in most cases required by law.[49] Accordingly, low-altitude airspace in particular has not truly served as a publically *and* legally accessible space, or "splice."

Drone technology, by contrast, is more widely accessible in terms of cost, and increasingly, with respect to operability. The fact that it is unmanned also permits the production of smaller, lighter weight and more nimble technology which, compared to manned aircraft, can access a wider array of spaces.[50] And, as the technologies permitting longer flight-times improve, so too will the public's ability to use drones for a wider range of applications. Accordingly, a segment of airspace that has long been legally presumed as 'public' has only recently become physically accessible by members of the public. This new reality is, paradoxically, raising new questions about the legal designation of this space.

These questions arise in large part because, while lawmakers and courts designated navigable airspace as a "public space", judicial decision-makers specifically declined to identify the lower boundary between public airspace and private land. Landowners maintain a property interest in the airspace above their land, up to an uncertain height – "at least as much of the space above the ground as [they] can occupy or use in connection with the land."[51] A specific legal boundary was perhaps unnecessary because at the time when these airspace property cases were decided, manned aircraft seldom entered into the low-altitude airspace above private property. And on the rare occasions when entry occurred, it could reasonably be addressed on a case-by-case basis.[52]

---

48 See e.g. Froomkin and Colangelo, *supra* at 25.

49 See e.g. *Canadian Aviation Regulations*, s. 602.13.

50 For further discussion see e.g. Kristen Thomasen, "Drones Aren't Planes" (May 20, 2016) Slate Future Tense, http://www.slate.com/articles/technology/future_tense/2016/05/why_policymakers_shouldn_t_think_of_drones_just_as_planes.html.

51 *Causby, supra* at 266. *Didow, supra* at paras 38-40 (adopting *Causby* and *Bernstein*). See also Gray, "Property in Thin Air", *supra*.

52 E.g. *Causby, supra* at 261-262.

Drone technology however is expected to operate predominantly within low-altitude airspace – below manned aircraft. At the time of writing, drone regulations in Canada require that the technology must not be flown more than 300 feet above the ground.[53] Similarly, in the United States the ceiling has been set at 400 feet.[54] This mandates that drones must be operated relatively closely to the ground.[55] As a result of the legal ceiling on drone airspace, the public/private property status of the low-altitude airspace in which the technology operates is becoming a more significant legal and spatial question. And, as law professor Troy Rule emphasizes, the "ambiguous airspace rights laws are now threatening to impede the growth of an important new industry."[56] These uncertain rights also call the "public" designation of low-altitude airspace into question.

Various calls have been made for clearer legal designation of low-altitude airspace as a public space, with the aim of facilitating drone use.[57] For instance, federal law could define a portion of low-altitude airspace as publicly navigable, which would free drone operators from the risk of legal actions for trespass within that space.[58] Amazon and Google have made public proposals outlining a vision of a horizontally stratified public airspace that could permit a variety of drone operations to occur simultaneously at different altitudes, facilitating much wider use of drones, in-

53 See e.g. *Canadian Aviation Regulations*, ss. 602.41 and 602.45; Transport Canada, "Flying your drone safely and legally" (December 21, 2016) at: https://www.tc.gc.ca/eng/civilaviation/opssvs/flying-drone-safely-legally.html.

54 Federal Aviation Administration, "Operation and Certification of Small Unmanned Aircraft Systems" (2016), https://www.faa.gov/uas/media/RIN_2120-AJ60_Clean_Signed.pdf. FAA, "Advisory Circular 91-57A, Model Aircraft Operating Standards" (2015), http://www.faa.gov/ documentLibrary/media/AdvisoryCircular/AC_91-57A.pdf.

55 Currently, drone operations are also fairly restricted over people and buildings. However, as the safety of the technology progresses this may be expected to change. See e.g. Transport Canada, "Flying your Drone Safely and Legally" accessed January 2017: https://www.tc.gc.ca/eng/civilaviation/opssvs/flying-drone-safely-legally.html.

56 Rule, "Airspace in an Age of Drones" at 208.

57 Bradley L Garrett and Adam Fish, "Attack on the Drones: The Creeping Privatisation of our Urban Airspace" (December 12, 2016) The Guardian, https://www.theguardian.com/cities/2016/dec/12/attack-drones-privatisation-urban-airspace.

58 Froomkin and Colangelo at 26.

cluding for small parcel delivery.[59] Professor Troy Rule has also suggested municipal or local drone ordinances to permit greater local control over where and when drones can fly.[60]

Notably, calls to maintain a legally designated public airspace for drone use are often driven by commercial concerns, as was the case for commercial aviation. While beyond the scope of this chapter, further consideration of the power dynamics guiding ground- and air-space rights is certainly warranted. Meanwhile, while the splice, and experience, of private property may continue to change if owners are no longer able to exclude potentially unwanted technology from above their land,[61] in the event that low-altitude airspace is more clearly designated as legally public, it will be increasingly necessary to question the genuine public accessibility of this space. In this vein, the next section examines some further considerations for an increasingly "public" airspace.

---

59 Amazon Prime Air, "Determining Safe Access with a Best Equipped, Best-Served Model for Small Unmanned Aircraft Systems" at https://images-na.ssl-images-amazon.com/images/G/01/112715/download/Amazon_Determining_Safe_Access_with_a_Best-Equipped_Best-Served_Model_for_sUAS.pdf; Amazon Prime Air, "Revising the Airspace Model for the Safe Integration of Small Unmanned Aircraft Systems" at https://images-na.ssl-images-amazon.com/images/G/01/112715/download/Amazon_Revising_the_Airspace_Model_for_the_Safe_Integration_of_sUAS.pdf; Google "Project Wing" at https://x.company/wing/.

60 See e.g. Troy Rule, "Drone Zoning" (2016) 95 North Carolina Law Review 133; Rule, "Airspace in an Age of Drones", *supra.*

61 Consider e.g. *Florida v Riley*, 488 US 445 (1989) in which the U.S. Supreme Court affirmed that one cannot expect privacy on private property visible from public airspace. In that case a helicopter was hovering 400 feet above the accused's fenced property. See also: *Dow; Ciraolo*; *Tessling*, *supra.* Though, as Rule suggests, there are ways in which drone regulations can potentially address this concern: Rule, "Drone Zoning", *supra.* However, in both Canada and the United States, to date, airspace restrictions have been imposed predominantly or exclusively by federal safety agencies, an approach which has been called into question. See in particular, Margot E. Kaminski, "Drone Federalism: Civilian Drones and the Things they Carry" (2013) 4 California Law Review 57-74; Rule "Drone Zoning", *supra.*

III. The "Public" Nature of Drone-Accessible Airspace

In both Canada and the United States, federal legislation delegates the management of the airspace to government agencies – Transport Canada and the Federal Aviation Administration respectively.[62] While the navigable airspace is legally understood as "public," access to it is heavily regulated by these agencies for the purpose of ensuring its safe use. For instance, airspace is segmented into classes, access to which is predicated on various considerations including the type of operation at issue, the aircraft entering the space, and various communications requirements.[63] Training, licencing and other pre-requisites are also required for pilots.[64]

Drone use in the airspace is regulated as well, to a heavier or lighter degree depending the drone operator. At the time of writing, individuals flying drones for recreation are subject to considerably fewer rules than commercial operators – though this appears to be shifting.[65] As noted above, both nations limit the height to which a drone can be flown. Drones similarly should not be flown within 9 kilometers of an airport or aerodrome in Canada, and in the United States only with permission.[66] The regulators currently also limit the use of drones beyond the line of sight, at night and

62 In Canada see: *Aeronautics Act*, R.S., 1985, c A-2; *Canadian Aviation Regulations*, SOR/96-433. In the U.S. see *Civil Aeronautics Act*, *supra*; *Federal Aviation Act of 1958*, enacted as Public Law 85-726, 72 Stat. 737, 49 USCA (Supp. 1958). The jurisdiction of these agencies to manage the airspace immediately above private property has come into question, for instance in the United States as a result of the FAA assertion that it has jurisdiction over everything in airspace. For a further discussion of this issue see: Rule, "Drone Zoning" supra at 16-17; Federal Aviation Administration, "Busting Myths about the FAA and Unmanned Aircraft" accessed January 2017: https://www.faa.gov/news/updates/?newsId=76240 ("The FAA is responsible for the safety of U.S. airspace from the ground up.").

63 See e.g. *Canadian Aviation Regulations*, s. 600.1 (Subpart 1 – Airspace).

64 See e.g. *Canadian Aviation Regulations*, s. 400.1. See also, Transport Canada, Notice of Proposed Amendment 2015-012 (Unmanned Air Vehicles). Executive summary available at: http://wwwapps.tc.gc.ca/Saf-Sec-Sur/2/NPA-APM/doc.aspx?id=10519.

65 See e.g. CARAC, "Notice of Proposed Amendment – Unmanned Air Vehicles" (May 28, 2015) CARAC Activity Reporting Notice #2015-12, accessible at: http://wwwapps.tc.gc.ca/Saf-Sec-Sur/2/NPA-APM/actr.aspx?id=17&aType=1&lang=eng.

66 See e.g. *Canadian Aviation Regulations*, ss. 602.41 and 602.45; Transport Canada, "Flying your drone safely and legally" (December 21, 2016) at: https://www.tc.gc.ca/eng/civilaviation/opssvs/flying-drone-safely-legally.html; Federal

in poor weather, though some of these requirements appear likely to relax in the future.[67] Additional operator requirements are in place, including registration, licencing and insurance.[68] These federal safety agencies also have the delegated power to prohibit or restrict entry into specific airspaces, again for the purpose of safety and security.[69] These rules are in place with the aim of ensuring the safety of airspace as well as safety on the ground.

These laws set out complex boundaries for where drones can lawfully operate, who can access that space, and for what purpose. The low-altitude airspace is consequently not simply open to the public, in the sense that access is unregulated. But, presumptively, any member of the public should be able to lawfully access the airspace with a drone assuming they fulfill the defined pre-requisites, much like the higher altitude airspace reserved for manned aviation. Accordingly, these safety rules do not on their face inherently alter the "public" nature of airspace.

However, legal geography also prompts us to question *who* adopts applicable regulations and makes the legal determination that a space is private or public, how, and for what purpose.[70] With the advent of an airspace technology like the drone, which engages numerous social values beyond safety, including freedom of expression, the almost exclusive prioritization of safety will increasingly need to give way to deeper concerns about the accessibility and use of this space.[71]

Aviation Administration, "Operation and Certification of Small Unmanned Aircraft Systems" (2016), *supra.*

67 Transport Canada, "Flying your Drone Safely", *supra.*

68 Rules apply differently to commercial and recreational users, with much stricter regulation of commercial operators though the distinction is beginning to diminish, per: *Canadian Aviation Regulations*, ss. 602.41 and 602.45. Transport Canada has proposed new rules, which are intended to come into effect in spring 2017, in which regulation for drones below a certain weight will be standardized regardless of the operator. Transport Canada, Notice of Proposed Amendment 2015-012 (Unmanned Air Vehicles). Executive summary available at: http://wwwapps.tc.gc.ca/Saf-Sec-Sur/2/NPA-APM/doc.aspx?id=10519. See also Federal Aviation Administration, "Getting Started" accessed January 2017: https://www.faa.gov/uas/getting_started/.

69 See e.g. various provisions under *Canadian Aviation Regulations*, s. 600.1.

70 Layard, "Freedom of Expression", *supra.*

71 Ruppert, "Rights to Public Space", *supra*; Layard, "Freedom of Expression", *supra.* See also Shayna Gersher, "Canada's Domestic Regulatory Framework for RPAS: A Call for Public Deliberation" (2014) 2 Journal of Unmanned Vehicle

Recent examples of the U.S. Federal Aviation Administration designating no fly zones over political protests, first in Ferguson, Missouri[72] and later over the North Dakota Access Pipeline protests (#NoDAPL) near Standing Rock Sioux Tribe lands call attention to this issue.[73] Drone footage from the #NoDAPL protests, for instance, captured clashes between protesters and police, serving to underscore allegations of policy brutality at the site.[74] These no fly zones, adopted by the airspace *safety* agency, had the effect of prohibiting members of the public from making use of the (purportedly public) airspace above contested ground space to document activities, in the public interest. While the FAA stated that journalists could apply for exemptions to the North Dakota flight restriction, the agency only ever granted one such exemption – to a non-local, non-Indigenous journalist.[75] Accordingly, the space above the land, which has borne a legal history purporting public access, was tightly restricted, suggesting a more controlled and political reality than previous jurisprudence would suggest. The federal government restriction of access to and use of airspace over Indigenous lands is more problematic when properly placed within the colonial context and history of these spaces – an issue which merits further analysis and consideration, as regulators continue to legally conceptualize this airspace.

Systems 1-4, highlighting the lack of transparency and public input in the Canadian regulatory framework, calling for greater public involvement in the regulation of drones.

72 "ACLU Letter to FAA Protesting No-Fly Zone for Media in Ferguson" (November 4, 2014) written by Anthony E. Rothert and Lee Rowland, ALCU at: https://www.aclu.org/other/aclu-letter-faa-protesting-no-fly-zone-media-ferguson.

73 Vera Eidelman, "FAA Helps Police Suppress Reporting From Dakota Pipeline Protests" (December 19, 2016) ACLU at: https://www.aclu.org/blog/free-future/faa-helps-police-suppress-reporting-dakota-pipeline-protests..

74 Saba Hamedy, "Drone Footage Shows Recent Clashes at #NoDAPL Protest in Standing Rock" (November 27, 2016) Mashable at: http://mashable.com/2016/11/27/standing-rock-no-dapl-protest-drone-footage/#gbqROX.UuSqs; see also: "Drone Footage Captures Scene at Standing Rock" (November 26, 2016) NBC News at: http://www.nbcnews.com/video/drone-footage-captures-scene-at-standing-rock-817789507848 (showing the size of the gathering despite freezing weather).

75 See Eidelman, *supra.* See also Jason Koebler, "The Government is Using a No Fly Zone to Suppress Journalism at Standing Rock" (November 20, 2016), Motherboard, accessed January 2017: http://motherboard.vice.com/read/the-government-is-using-a-no-fly-zone-to-suppress-journalism-at-standing-rock.

The restriction of access to airspace, particularly for expressive purposes over public ground spaces serving as sites of expressive activities, is both legally and spatially problematic. Now that members of the public *can* actually access and utilize this "public" space, access is eliminated through restrictive regulation.[76] As lawmakers and courts clarify the legal designation of the now more publically accessible low-altitude airspace, the democratic values that inform public land-spaces should similarly inform drone law.[77]

IV. Considerations for Public Space on the Ground

Building upon the above, if further access is granted to low-altitude airspace for a wider range of drone activities, this will also have consequences for individuals and their experiences on the public ground below. Some potentially positive consequences have already been mentioned, in particular that drone technology offers a means of accessing and sharing important information in the public interest. But, in different circumstances, this technological capability might also have a negative impact on social experiences within the "splice" of public ground space. Perhaps one of the most immediate, and legally and socially significant, consequences will be with regard to privacy, and whether individuals or communities have or ought to have legal recourse to resist particular unwanted uses of the technology.[78] This section briefly identifies some of the potential

76 See also, Don Mitchell, "The Annihilation of Space by Law: The Roots and Implications of Anti-Homeless Laws in the United States" in Nicholas Blomley, David Delaney, and Richard T. Ford (eds), *The Legal Geographies Reader* (Oxford: Blackwell Publishers Ltd, 2001) at 6.

77 See Ruppert, "Rights to Public Space", *supra*; Nicholas Blomley, "Introduction" in *The Legal Geographies Reader*, *supra*, at 3.

78 For instance in Canada see C. Bracken-Roche, D. Lyon, M. Mansour, A. Molnar, A. Saulnier, and S. Thompson, "Surveillance Drones: Privacy Implications of the Spread of Unmanned Aerial Vehicles (UAVs) in Canada" (June 2014) *Report for the Office of the Privacy Commissioner of Canada,* available at: http://www.sscqueens.org/sites/default/files/Surveillance_Drones_Report.pdf ["Surveillance Drones"]; and Paul D.M. Holden, "Flying Robots and Privacy in Canada" (forthcoming) Canadian Journal of Law and Technology ["Flying Robots"], which consider the application of federal and provincial privacy statutes, tort law (Holden) and briefly consider constitutional protections (Bracken-Roche et al), and determine that while these laws apply to drones, there are challenges both in the

ground-space implications of greater public access to the airspace mediated through drones.[79]

For instance, the aerial nature of drone technology permits an operator to access potentially unexpected vantage points of the ground below, compared to ground-based or stationary mechanisms of observation (such as CCTV and cell phone video). Since there is no human on board, drone technology can also be smaller than a manned aircraft, and operated in areas too dangerous or difficult for manned aircraft to access – where aerial observation might not be expected or desired. This unmanned aerial nature also makes the technology well suited to longer-term monitoring and tracking, compared to a person (or manned aircraft) attempting to do the same. Furthermore, as previously discussed, since drones are relatively inexpensive (compared to manned aircraft) and can, at present, be sold to anyone, the technology can be put to wide-scale use by a variety of actors – this can have great value in terms of use of public airspace, but can also result in frequent ground-space encounters with the technology, some of which might be unwanted.[80] Additionally, the fact that the drone, and any information-gathering payloads, operates detached from the human pilot raises additional challenges for privacy and accountability. For example, an individual may feel that a drone has invaded her privacy, but if she cannot identify the operator – because the pilot is located at a distance, or the drone is operating autonomously – then she may not know whom to pursue, or through which legal mechanism.[81] The drone's detachment from

practical application of the law, and with the uncertain scope of protection. See also: Ann Cavoukian, *Privacy and Drones: Unmanned Aerial Vehicles* (Toronto: Information & Privacy Commissioner Ontario, 2012). But see John Villasenor, "Observations from Above: Unmanned Aircraft Systems and Privacy" (2013) 36 Harvard Journal of Law & Public Policy 457 who suggests that the gaps in privacy protection (in the U.S. context) are not as significant as many authors suggest.

79 Which I have also argued in: Kristen Thomasen, "Beyond Airspace Safety: Feminist Perspectives on Drone Regulation and Privacy in Public" (in progress), available at: http://www.werobot2017.com/wp-content/uploads/2017/03/Thomasen-Feminist-Perspectives-on-Drone-Regulation-We-Robot-1.pdf ["Feminist Perspectives"]

80 See also Thomasen, "Feminist Perspectives". And therefore consider for instance: Woodrow Hartzog and Evan Selinger "Surveillance as Loss of Obscurity" (2015) 72 Washington and Lee Law Review 1343.

81 See e.g. Froomkin and Colangelo, "Self-Defense Against Robots and Drones", *supra*.

the pilot can accordingly disempower the observed; she has no immediate way to gain more information about the drone's operation or operator.[82]

These features and capabilities, among others, can combine to cause a 'panoptic' chilling effect on individuals on the ground below.[83] As Professor Roger Clarke explains, "drones actually bring back the sense of physical superiority of the observation-point over ground-dwelling individuals."[84] Professor Ryan Calo has described drones as representing "the cold, technological embodiment of observation."[85] Accordingly, drone technology can not only collect enormous amounts of information from the ground below (for potentially positive or detrimental uses) but the technology also *feels* invasive.[86]

By virtue of this physical, aerial nature, drones can impinge on what Clarke has called "behavioural" privacy. This is an understanding of privacy that is concerned with the "freedom of the individual to behave as they wish, without undue interference from others."[87] A technology that en-

---

82 See also, Thomasen "Feminist Perspectives". This could also result in individuals taking dangerous self-help measures in response to drones. As Froomkin and Colangelo, *supra* explain at p. 33: "the less the victim knows about the robot spy, or suspected spy, the more that its surveillance is likely to seem a threat. And the more that the surveillance seems a threat, the more that the victim will seek not just a judicial remedy – uncertain, likely time-consuming and costly, and probably much too late to undo the harm – the more that the victim will seek a self-help remedy." Uncertainty about a drone's mission and owner played a significant role in an encounter leading to one of the first U.S. criminal cases involving a drone (see e.g. http://www.slate.com/blogs/future_tense/2015/10/28/case_against_william_merideth_for_shooting_down_a_drone_is_dismissed.html). In that case, Mr. Meredith, who felt his privacy had been invaded, shot the drone down.

83 Clarke, *supra* at 289-90.

84 *Ibid* at 288.

85 Ryan Calo, "The Drone as Privacy Catalyst" (2011) 64 Stan L Rev Online 29 at 34.

86 The breadth of media stories involving individuals fearing drones, shooting at drones, or otherwise seeking to disable them, anecdotally suggests that the wider use of the technology could continue to make individuals uncomfortable – this is both a civil rights issue and a commercial issue, as it could hinder broader acceptance and use of the technology. See also Thomasen, "Feminist Perspectives".

87 Clarke, *supra* at 287.

gages one's behavioural privacy can impinge upon one's autonomy and freedom.[88]

Therefore, widespread drone use of the sort envisioned by for instance Google or Amazon, might bring many social benefits through greater use of airspace, but could also negatively impact upon public activities and the enjoyment of public ground spaces – at least in so far as individuals have uncertain protection or control against unwanted observations or invasions.[89] Importantly, such an outcome could risk undermining the values and use of public (ground) space, paradoxically, in the interest of expanding the beneficial uses of public (air) space.

While serving as a powerful tool for expression and commerce, among other things, drones are also capable of overcoming various physical and normative boundaries that can otherwise serve to protect and shape interactions between individuals on the ground. If (or as) drones laws permit more widespread use of the technology in the airspace, these laws could have the unintended effect of re-shaping lived experiences on public and publically-accessible land.[90] Lawmakers therefore ought to consider how to replace these normative and physical barriers with legal protections in public spaces. Alternatively, despite their many social benefits, the increasing accessibility of public airspace through drone technology might also undermine some of the previously assumed boundaries that protected individual interests and experiences in public space on the ground, paradoxically undermining the 'public' nature of public ground space.

D. Conclusion

Drone technology stands to (re)shape the legal conceptualization of low-altitude airspace, by permitting members of the public to more widely access this space and use it for a variety of activities. This new accessibility to a previously relatively unused space places drone technology at a chal-

88 The Supreme Court of Canada has identified these as significant values enshrined in the s. 8 *Canadian Charter of Rights and Freedoms* protection against unreasonable search and seizure. See e.g. *R v Spencer*, 2014 SCC 43 at para. 15.

89 See in particular: Bracken-Roche et al., "Surveillance Drones" and Holden, "Flying Robots".

90 As well as on private land, however as discussed, landowners have a legal right to exclude the technology up to a particular (though currently ill-defined) height above their private property.

lenging nexus of law and space, not unlike manned aircraft previously. However, given the wider range of applications to which a drone can be put, and the broader population to whom the technology is accessible, the social dynamics and values implicated by increasing drone use are considerably more complex.

This paper has accordingly drawn from concepts within the field of Law and Geography to consider some of the spatial and social implications of drone laws. Ultimately, Law and Geography provides a critical lens through the legal, spatial and social implications of a technology that stands to change the nature of a previously under-utilized space may be analyzed.

Google Cars, Software Agents, Autonomous Weapons Systems – New Challenges for Criminal Law?

*Susanne Beck,**
University of Hannover

A. Introduction

The Internet is supposed to enable its users to find out about any subject without any restriction, to communicate with other users without being observed by third parties, to work on their own computers without outside surveillance. Whilst this desire for free use of the Internet may be understandable, there is also a large degree of consensus that the Internet cannot and must not be free of the rule of law. It is not just the fact that free access to information and the free use of the existing technical possibilities would be hard to guarantee without the protection of the law; also, some spheres of communication need protection against outside intervention, some content needs to be prohibited, and some statements need to have their truthfulness established in law. For this reason, the national legislature and many international and supranational institutions have been working for many years to impose legal constraints.

However, legal control of the Internet, particularly in terms of criminal law, faces substantial difficulties: many users act across borders, anonymously, without the social control which is imposed in real life by the presence of others. For this reason, the doubts about the suitability of criminal law in particular to safeguard protected interests on the Internet have never disappeared. Also, society attaches different weightings to the respective interests and takes differing views of the degree of protection

---

* The original text was published by Susanne Beck, Google-Cars, Software-Agents, Autonome Waffensysteme – neue Herausforderungen für das Strafrecht?, in: Beck, Susanne/Meier, Bernd-Dieter, Momsen, Carsten (ed.): Cybercrime und Cyberinvestigations. Neue Herausforderungen der Digitalisierung für Strafrecht, Strafprozessrecht und Kriminologie, Baden-Baden 2015, p. 9-33.

required by certain endangered interests, such as honour, law and order, or the undisturbed development of minors.[1]

The following considerations are to focus on the way criminal law deals with networked machines. Here, I will concentrate on electronic agents which act on the Internet or use the Internet as a source of information, are networked via the Internet or controlled by a person via the Internet. I will start by presenting the various possible situations and study their potential conflicts in some detail. I will then analyse the possibilities offered by criminal law to resolve conflicts, illustrate its limitations, and discuss possible changes.

B. Semi- and Fully Autonomous Machines and the Internet

The Internet has long since ceased to serve only for the purpose of communication between people – it is now also possible for a person to control machines and communicate with them via the Internet; in some cases, machines actually communicate directly with one another.

I. Networking of People with Embodied Machines

As an example of interactions between people and machines in or via cyberspace, let us first cite "Giraff",[2] a robot currently in the test phase which can serve as an example for the research field of "ambient assisted living" (AAL). AAL means that the environment of older people or people in need of care – i.e. particularly their dwelling – is enhanced by machinery which allows them to pursue a largely independent life in a familiar setting.[3] For example, the fridge can work out itself when certain foodstuffs are running out and order them directly online, or the mobile phone

---

1 On the protection of honour and the problem of anonymity on the Internet, cf. *Heckmann*, Persönlichkeitsschutz im Internet – Anonymität der IT-Nutzung und permanente Datenverknüpfung als Herausforderung für Ehrschutz und Profilschutz, NJW 2012, p. 2631 *et seq*.

2 http://www.giraff.org/?lang=en (18 May 2015).

3 Further information about research into ambient assisted living can be found in *Georgieff*, Ambient Assisted Living – Marktpotenziale IT-unterstützter Pflege für ein selbstbestimmtes Altern, http://www.fazit-forschung.de/fileadmin/_fazit-forschung/downloads/FAZIT-Schriftenreihe_Band_17.pdf (18 May 2015).

can remind the user to take his medicine or to drink enough fluids. Further to this, sensors in the dwelling can work out if the resident has fallen over or is staying in bed so long that he is probably ill. All of these actions would be categorised as autonomous acts by the machines – with "autonomous" being understood here in a wider sense, i.e. that the machine has a scope for decision-making which is not influenced by people.[4] I shall come back in a moment to further examples of autonomously active machines, but would like to start by addressing machines networked with people.

The "Giraff" robot is a remote-controlled computer screen on wheels which can move through a person's dwelling or can be controlled by someone in that dwelling and can enable conversations to take place online with relatives, pastoral carers or doctors, and which enables a doctor to decide whether an ambulance is needed in an emergency.[5]

Giraff is just one example of machines – including partially autonomous machines – which are remotely controlled via the Internet. It is also feasible (and is already happening in some cases) that vehicles, medical appliances, drones and other weapons systems can be remotely controlled in this way.[6] Through the Internet, the person exercising control receives the information which the machine has picked up through its sensors, and on this basis issues orders which the machine subsequently implements. Partial autonomy is feasible given the way the information is obtained, selected and even evaluated. The specific implementation of a rather abstract command can also in principle be defined and undertaken by the machine. But in this situation, the key decisions are taken by a human being.

II. Networking of Autonomous Machines

The situation is different in the case of autonomous machines. The rules on their actions are very general; for example, electronic agents can independently activate selling and buying on the stock exchange, merely fol-

---

4 *Christaller et al.*, Robotik, 2001, p. 18 *et seq.*; cf. also *Beck*, Roboter, Cyborgs und das Recht – von der Fiktion zur Realität, in: Spranger (ed.), Aktuelle Herausforderungen der Life Sciences, 2010, p. 101.

5 http://www.giraff.org/about-giraff/?lang=en (18 May 2015).

6 Cf. *Beck* (footnote 4), p. 96; *Christaller et al.* (footnote 4), p. 46 *et seq.*

lowing general targets of profit maximisation and risk tolerance.[7] Electronic agents are non-embodied, purely virtually active software programs. The conclusion of contracts is just one area in which they can operate – beyond that, many Wikipedia articles and other sources of information already use such agents; in future, they may even take over some communication between people, such as writing and responding to emails on their own.

Embodied machines could also act autonomously.[8] For example, a vehicle could travel from A to B without human intervention, the afore-mentioned fridge could order food based on the information it has collected about its user's preferences, a weapons system could monitor a certain area without the involvement of a human commander and injure or even kill people assessed as enemies. These are just a few examples of possible areas for the deployment of autonomous machines; from education to care for the elderly, from industrial manufacturing to medical treatment, from fitness training to sexuality – they can be used in future in all of these and other areas of life.

Whilst software agents have to rely on the Internet in order to act, since they only exist virtually, this is not necessarily the case with embodied autonomous machines. These could merely be programmed once at the beginning, and then act in line with the originally prescribed objectives and capabilities, the information collected by the sensors, and learnt or taught modes of conduct.[9] In practice, however, most of these machines will also be connected to the Internet or with one another in order to permit access to the information stored there, to external controls which may occasionally be required, and for mutual coordination.[10] For example, a self-driving vehicle can respond more appropriately to traffic if it learns in advance from the other vehicles about their planned routes, speeds, etc. The fridge also might place a different order if it is informed by the bodily function robot that the person is currently suffering from particularly high blood

7 Cf. *Neuhäuser*, Roboter und moralische Verantwortung, in: Hilgendorf (ed.), Robotik im Kontext von Recht und Moral, 2014, p. 270.

8 Regarding the full automation of a vehicle *Both*, Hands-Free Driving? Automatisiertes Fahren und Mensch-Maschine Interaktion, in: Hilgendorf (footnote 7), p. 172 *et seq*.

9 *Neuhäuser* (footnote 7), p. 270.

10 *Gruber*, Zumutung und Zumutbarkeit von Verantwortung in Mensch-Maschine-Assoziationen, in: Hilgendorf/Günther (ed.), Robotik und Gesetzgebung, 2013, p. 126.

pressure or insulin levels – perhaps it would even find out online what foods are better and worse in this situation. And autonomous weapons systems could tell one another which nation they belong to and thus render a complex sensory process to distinguish friend and foe redundant.

C. Potential for Conflict in this Development

It is not really a surprise that the increased use of machines in our lives offers considerable potential for conflict.

I. With Human Involvement

On the one hand, it is feasible that a person will deliberately use a machine to violate others' interests. The control of a machine via the Internet may even enable the user to commit crimes with no risk of injury or discovery. For example, a remotely controlled armed robot could rob a bank. Rather more frequently than such serious crimes, it might happen in future that a relative, for example, uses the service robot of the older relative to move around the dwelling without their knowledge and to monitor them, thus – depending on the situation and legal assessment – committing the offence of *Hausfriedensbruch* (here akin to criminal trespass) or violating data protection rules.[11] These and similar violations of others' interests become possible due to the fact that the user does not need to be at the location, but can control the machine via the Internet.[12]

The networking of people and machines is particularly dangerous in that it can probably result in unintended violations of protected interests on a regular basis.[13] For example, it is feasible that a service robot used for a medical or pastoral meeting needs to be serviced by a technician, and that the technician can eavesdrop on the confidential content. A remotely controlled vehicle could damage other people's property, or information saved about someone's own trips and others' trips could be open to unauthorised access. The decision to destroy a certain target using a remotely

11 Regarding data protection issues in this context: *Bräutigam/Klindt*, Industrie 4.0, das Internet der Dinge und das Recht, NJW 2015, p. 1139 *et seq.*

12 Cf. *Neuhäuser* (footnote 7), p. 270.

13 Cf. *Gruber* (footnote 10), p. 126.

controlled drone could be wrong, due to false information or delays in transmission, and result in the killing of civilians. These few examples show the considerable potential for conflict in the networking of people and machines.

II. Without Human Involvement

The use of machines which take decisions independently of people itself creates substantial risks for the interests of the users and third parties. Firstly, at least at present, it seems likely that decisions taken by machines will be substantially prone to error.[14] Secondly, the machines lack qualities which – in addition to rationality and a large amount of information – play an important role in human interactions, such as empathy and creativity.[15] This can also mean that decisions are deemed inadequate and erroneous from the human perspective.

The latter aspect is linked with another problem of autonomous machines which is already the subject of an ethical and legal debate: the "responsibility gap".[16] This is because the people who assess the machine's decision, or are even affected by it, are not only confronted with the lack of empathy, but also with the fact that the decision is taken without another human being taking responsibility for it. After all, the point of autonomous machines is that they relieve people of the burden of taking certain decisions – when they park their car, buy goods in online auctions, or even in war. This relief is only effective if the person is not subsequently liable for all the machine's wrong or not humanly comprehensible decisions. Further to this, it is virtually impossible to ascribe a wrong decision to a single participant without further investigation: given the fact that many wrong decisions taken by the machine are also likely to be due to

14 Examples for such wrong decisions can be found in *Müller-Hengstenberg/Kirn*, Intelligente (Software-)Agenten: Eine neue Herausforderung unseres Rechtssystems – Rechtliche Konsequenzen der „Verselbstständigung" technischer Systeme, MMR 2014, p. 309 *et seq*.

15 http://www.spiegel.de/wissenschaft/technik/autonome-waffen-roboter-sollen-krieg-ethischer-machen-a-977614.html (18 May 2015). cf. also https://netzpolitik.org/2012/kriegsroboter-und-drohnen-human-rights-watch-fordert-verbot-von-killer-robots/ (18 May 2015).

16 *Matthias*, Automaten als Träger von Rechten. Plädoyer für eine Gesetzesänderung, 2nd ed. 2010, p. 13 *et seq*.

programming errors or material failures, it hardly seems reasonable to make the user comprehensively liable.[17] Also, autonomous machines are so complex that in almost all cases it will be impossible to pinpoint the specific cause of the wrong decision.[18] For this reason, and because the user at least decides on the deployment of the machine, and may be able to train it and orient it to certain situations, a comprehensive liability for the programmer, manufacturer or seller is equally hard to justify.

This means it is not actually possible to identify a responsible person in such situations for the decisions which we normally regard as interpersonal interactions and for which we normally hold someone responsible in our social practice. Thus, those affected frequently respond with the need to hold someone to account, whether it is for damage to a car after a tailgating incident or even for the killing of a person by the opponent's weapon in a war situation. Possibly – at least, this is being discussed in the context of autonomous weapons systems – they even have a right to have this decision taken by another human being who has to live with that decision.[19] Since, as mentioned, autonomous machines are designed to relieve the burden on people, the problem of the responsibility gap cannot be resolved by designing the machines differently.

These few considerations show that the existing moral and legal concepts cannot simply be applied to the individual autonomous machine. The following section discusses whether it is possible to adapt the legal system to a certain extent and thus to make the problem manageable. For now, we can state that even non-networked autonomous machines could cause substantial social conflicts.

The networking of the machines makes the difficulties even greater. And problems can also derive from the use of the Internet: connections can be unstable or hacked; the information accessed can be false or even

---

17 Regarding the attribution of declarations of intent and liability problems under civil law: *Bräutigam/Klindt* (footnote 11), p. 1137 *et seq.*

18 Cf. *Beck*, Roboter und Cyborgs – erobern sie unsere Welt? Eine nicht futurologische Einführung in die Thematik, in: Beck (ed.), Jenseits von Mensch und Maschine, 2012, p. 15.

19 *Sharkey* in particular points in this direction in a Spiegelonline report entitled "Autonome Waffen: Wie Roboter den Krieg menschlicher machen sollen" of 26 June 2014. *Sharkey* takes the view that machines should fundamentally not be able to decide whether a human being lives or dies. Delegating such decisions to robots is deeply inhuman, http://www.spiegel.de/wissenschaft/technik/autonome-waffen-roboter-sollen-krieg-ethischer-machen-a-977614.html (18 May 2015).

deliberately falsified.[20] False evaluations or decisions by a machine can be disseminated through networking with other machines. Finally, the linkage itself can result in damage because it creates cumulative effects which cannot be generated by individual machines.[21]

An example of such a chain reaction is the stock exchange crashes[22] caused by the interaction of innumerable electronic agents – all programmed to sell shares under certain conditions – within the space of a few seconds. It is true that most of the severe falls in the stock prices were subsequently reversed, but this does clearly show how quickly and uncontrollably the networking of machines can result in damage to others' interests or assets, and can even shake economic systems. The interaction of several machines, each of which decides autonomously, has learnt from its own experience and pursues its own aims, exponentially increases the unpredictability and uncontrollability of their decisions.

So the potential for conflict ultimately rests in the fact that the problems of the Internet combine with the difficulties deriving from the automation of machines. The result is areas which are unfettered by responsibility and which entail the risk of substantial dangers.

D. Possible Solutions in Current Criminal Law

If one views criminal law as a means of conflict resolution, it can be deployed where an individual offender culpably violates or endangers an interest protected by law.[23] Whether and to what extent a role is played by

---

20 *Gruber*, Was spricht gegen Maschinenrechte?, in: Gruber/Bung/Ziemann (ed.), Autonome Automaten. Künstliche Körper und artifizielle Agenten in der technisierten Gesellschaft, 2014, p. 195.

21 Cf. *Müller-Hengstenberg/Kirn* (footnote 14), p. 310.

22 Such as "Black Monday" of 19 October 1987 or the "Flash Crash" of 6 May 2010.

23 Regarding the principle of culpability cf. *Radtke*, in: Joecks/Miebach (ed.), Münchener Kommentar zum StGB, vol. 2, 2nd edition. 2012, before section 38 *et seq*. para. 14.

retribution[24] or prevention,[25] and to what extent the deployment of criminal law mainly serves to communicate the reprehensibility of an action and the need to protect certain interests to the offender, the victim and society, varies according to which theory of punishment is applied.[26] The focus here is not on the plausibility of the various theories; the internal systemic logic of criminal law is primarily of significance in the following arguments. This is based, it should be pointed out, on the premise that criminal law should be understood in a constructivist manner,[27] that criminal law has a certain force as a means of communication, and that therefore criminal law is useful and effective only if it embraces the conflict in an appropriate manner, attributes responsibility in a reasonable way, and corresponds to the (presumed) social values as far as possible.[28]

The following section describes, before any assessment is possible, what possible solutions current criminal law offers for the handling of networked machines.

I. With Human Involvement

1. Machines as Weapons, Tools or Instruments for Committing Crimes

Irrespective of the specific crime committed, be it the case of the remote control of an armed robot, or particularly when a machine is used for certain illegal purposes (which might be bank robbery, but also illegal transactions via electronic agents) it could be discussed whether the action is

---

24 The absolute theories of punishment see the purpose of punishment in retribution and the restoration of justice, for more details cf. *Joecks*, in: Joecks/Miebach (ed.), Münchener Kommentar zum StGB, vol. 1, 2nd edition 2011, introduction to para. 48 *et seq*. with further citations.

25 In contrast, the relative theories of punishment see the purpose of punishment in the prevention of crime, cf. also in this respect *Joecks* (footnote 24), introduction to para. 56 *et seq*. with further citations.

26 An overview of the theories of the purpose of punishment can be found in *Joecks* (footnote 24), introduction to paras. 47-76; cf. also *Hörnle*, Straftheorien, 2011, p. 15 *et seq*.; on the history of theories of punishment *von Hippel*, Lehrbuch des Strafrechts, 1932, p. 56 *et seq*.

27 On the constructivist theory of crime *Hess/Scheerer*, Was ist Kriminalität? Skizze einer konstruktivistischen Kriminalitätstheorie, KrimJ 1997, p. 83 *et seq*.

28 *Hörnle* (footnote 26), p. 7 *et seq*.

actually undertaken by the controller/commander.[29] As long as this person actually directly controls the machine in real time, this is probably the case, even if the offender is not at the location. However, the fact that this is far from easy to answer in the case of autonomous machines is discussed in detail below.[30] The same goes for the question of when, in such cases, the crime commences, i.e. in particular when the threshold to the punishable attempt to commit a crime has been passed. This will also be discussed in detail, in this case in the section about the fully autonomous machines, as it is of greater relevance there.

It is also possible, irrespective of specific norms, to discuss the extent to which private citizens are permitted to protect themselves against robots, and to what extent robots can be used for protection. The first question depends on whether the robot's conduct is an "attack" within the meaning of Section 32 of the Criminal Code.[31] So far, only human actions have been covered by this term, whilst dangers deriving from objects could only justify an act of "necessity". However, if the object is remotely controlled, there is nothing in principle to argue against equivalency with a human attack; this solution is not least possible because the attacked person currently, due to the rarity of autonomous weapons, will presume that control is being exercised by a human being when a robot deliberately "attacks" – as already discussed, the assumption here would also be that a human action was taking place.

Whether the protection of one's own interests by an armed robot is admissible – e.g. if it can be moved around by remote control outside one's house and caused to fire at intruders[32] – cannot be answered in general terms. In principle, there is a far-reaching right to protect oneself against attackers,[33] so that as a rule there will be no concerns about such security measures as long as the corresponding public-law rules (Weapons Act)[34]

---

29 Cf. *Beck* (footnote 18), p. 15.

30 Cf. the comments under C. II.

31 Regarding the preconditions for an attack within the meaning of Section 32 of the Criminal Code, cf. *Perron*, in: Schönke/Schröder (ed.), StGB, 29th edition 2014, section 32 para. 3 *et seq.*

32 Cf. on this *Beck* (footnote 4), p. 108 *et seq.*

33 Regarding the range of rights to self-defence *Erb*, in: Joecks/Miebach (footnote 24), section 32 para 9.

34 Of 11 October 2002 (Federal Law Gazette I 2002, pp. 3970, 4592; Federal Law Gazette I 2003, p. 1957), most recently amended by Art. 4(65) of the Act of 7 August 2013 (Federal Law Gazette I p. 3154).

are complied with and the user verifies in each case whether the attack can be warded off by milder means and whether the deployment of the robot is actually appropriate.

With a view to the committing of specific crimes, there are initially no special problems in those cases in which the machine is deliberately used by remote control as a weapon or tool to commit a crime planned by people (bank robbery, murder, etc.) for which a person does not necessarily have to be at the location or for which a certain, precisely described action is required – depending on the crime, Sections 249, 239 or Sections 212, 223 of the Criminal Code apply. There can be virtually no doubt that a correspondingly armed robot is a weapon[35] or at least a dangerous tool.[36] Occasionally, it may also be necessary to decide whether an unarmed machine is a dangerous tool – but that will probably be deemed to be the case in many situations.

In contrast, it is much more difficult to categorise the cases in which the crime presupposes specific human conduct or presence at the location. For example, the movement of a service robot (making image and sound recordings) which has not been approved by the resident through his dwelling could be regarded as a criminal trespass within the meaning of Section 123 of the Criminal Code, as in terms of the protected interest (domiciliary rights)[37] and the perceived violation of the interest, it can certainly be regarded as being similar to the undesired presence of a person in the dwelling; in particular, the remotely controlled machine violates privacy and passes on intimate information to a person, and the machine could also attack and injure the resident.

At the same time, there is probably a consensus that the mere transmission of information from a dwelling – e.g. via a secretly installed camera or a directional microphone – does not represent a crime in terms of Sec-

35 The concept of a weapon in criminal law embraces every physical object which is suited, in terms of its objective design and condition at the time of the crime when used for its intended purpose, to cause considerable injuries, BGH NJW 2003, p. 1678.

36 A dangerous tool is defined as any movable object which due to its physical characteristics and the way it is used in a specific case is suited to cause substantial bodily harm, BGH NStZ 2010, p. 513.

37 *Ostendorf*, in: Kindhäuser/Neumann/Paeffgen (ed.), StGB, vol. 2, 4th edition 2013, section 123 para. 4 with further citations.

tion 123 of the Criminal Code.[38] Unlike the robot, the camera or the microphone do not usually move around; also, the data transmission takes place secretly. But these differences cannot be decisive for criminal liability, because whether or not the crime is noticed has no effect on whether or not a crime is committed pursuant to Section 123 of the Criminal Code.[39] And even if the (presumed?) independence of the machine can create a different sense of threat than a video camera – ultimately the existing criminal law fails to cover the situation due to the limits to the meaning of the wording, which is so important in this field of law. For an "entry" within the meaning of Section 123 of the Criminal Code, it is presumably necessary for a person to be at the location[40] and to deliberately enter the protected area.

Similar problems of interpretation arise in the questions as to whether the remote control of a robot can be qualified as "driving a vehicle"[41] within the meaning of Section 315c or 316 of the Criminal Code, or whether waiting for someone or following them via a machine could be "stalking"[42] within the meaning of Section 238 of the Criminal Code. Without being able to go into all of these questions in detail, it should be pointed out that there is a conflict here between teleological and grammatical interpretation – even if the legislature has not thought of such cases, they do threaten the protected interest in a similar way to a person on the spot; whether the wording can be interpreted to cover this without going too far would need to be discussed specifically for each clause. Overall, however, it is difficult to resolve these conflicts within the rules, as the rules require a certain mode of conduct. This is all the more true of autonomous machines which act without human involvement.

This does not imply that all such modes of conduct are not punishable. For example, if the robot records images or sound, this may be punishable pursuant to Sections 201, 201a of the Criminal Code. Here, the robot

---

38 Cf. for example *Ostendorf* (footnote 37), section 123 para. 26, who does qualify the brief entry to install eavesdropping devices – but by implication not the eavesdropping itself – as "entry" within the meaning of Section 123 of the Criminal Code.

39 *Sternberg-Lieben*, in: Schönke/Schröder (footnote 31), section 123 para. 11.

40 *Sternberg-Lieben* (footnote 39), section 123 para. 12.

41 *Sternberg-Lieben/Hecker*, in: Schönke/Schröder (footnote 31), section 316 para. 19.

42 *Lackner*, in: Lackner/Kühl (ed.), StGB, 28th edition 2014, section 238 para 3.

would need to be an "eavesdropping device"[43] within the meaning of Section 201(2) no. 1 of the Criminal Code, and if the device is deliberately steered to the vicinity of a conversation in order to eavesdrop on it, it is hard to deny that. If it is to be punishable pursuant to Section 201a of the Criminal Code, the transmission or creation of the images must also violate intimate privacy[44] – if the robot is deployed in a dwelling in order to record the behaviour of the resident, this will however usually be the case at least in terms of some of the images transmitted, since one feels unobserved in one's own dwelling and therefore many of the acts taking place there are likely to be deemed intimately private.

Finally, violations of data protection law are possible.[45] Here, it should only be mentioned as a marginal note that the storing of the image in conjunction with the possibility to use facial recognition software to obtain further data about the person could be punishable under Section 44 of the Federal Data Protection Act. In this regard, it is particularly questionable whether the streaming itself represents a transmission or use of the data. Without going into this argument in detail – we will also be hearing more about data protection on the Internet and thus also streaming here in a few weeks' time – it should be pointed out that this problem will be exacerbated in the future when the use of remotely controlled robots increases in the public and private arena. After all, the remote control is only possible if the surroundings are recorded and observed by the controller.

In terms of material criminal law, it should be stated at this point that an extended interpretation of at least some of the provisions and constructs of the Criminal Code can include the remote control of a robot by a person – but such an interpretation would be close to the limits offered by the wording in many cases.

2. *Duties of Care When Remotely Controlling a Robot*

Some damage which occurs when robots are remotely controlled could be covered by negligence penalties. The use of remotely controlled robots

---

43 *Graf*, in: Joecks/Miebach (ed.), Münchener Kommentar zum StGB, vol. 4, 2nd edition 2012, section 201 para. 32 *et seq.*

44 Cf. in this regard *Flechsig*, Schutz gegen Verletzung des höchstpersönlichen Lebensbereichs durch Bildaufnahmen, ZUM 2004, p. 607.

45 More details at *Bräutigam/Klindt* (footnote 11), p. 1139 *et seq.*

must of course comply with certain safety standards. At present, some standards are developing at international and national level (ISO and DIN standards)[46] for research and manufacturing in robotics. Even if the generation of these standards certainly merits criticism in terms of transparency and representative participation, there are ways to include them in criminal law without fully taking over the values expressed in them. In particular, if specific standards fail to protect certain interests – e.g. physical safety or privacy rights – then negligence would occur even though the standards had been met. In such cases, however, the assessment should embrace not only the protection of interests, but also the reprehensible nature of the specific action and the individual guilt of the person exercising remote control.[47] In particular, it is necessary to discuss how to handle the belief of those active in the field of robotics that compliance with the standards exempts them from all legal liability.[48] Since the robotics sector is only just starting out on an in-depth consideration of the ethical and legal consequences of their actions by working together with experts in standards and law, it seems unlikely that an awareness of the wrongful nature of actions will develop quickly. At the same time, of course, neither programmers and producers nor users deny that they are aware of the risks of such machines.

In particular, they generally know that the use of the Internet in particular entails additional risks: the control of a robot can result in damage for example because the Internet was down, because the transmission of what can be – e.g. in the case of Giraff – psychological and therapeutic discussions was inadequately protected, or because the involvement of a technician resulted in the disclosure of personal, or even secret, information within the meaning of Section 202 *et seq.* of the Criminal Code.[49]

But even if duties of care are violated, it is not always clear in such situations which of the participants is responsible for the violation of the protected interest. The distribution of responsibility, which will be discussed in detail below with relation to autonomous machines, is also of impor-

46 Regarding current standardisation projects in robotics cf. http://robotik.vdma.org/article/-/articleview/569335 (18 May 2015).

47 Cf. *Beck* (footnote 4), p. 108 *et seq.*

48 Regarding the exemption from liability where legal standards are complied with in the field of civil law: *Spindler*, Zivilrechtliche Fragen beim Einsatz von Robotern, in: Hilgendorf (footnote 7), p. 70 *et seq.*

49 *Kargl*, in: Kindhäuser/Neumann/Paeffgen (footnote 37), section 202 para. 2.

tance for criminal liability here, since there is a need to specifically name the person(s) responsible if they are to be held criminally liable. In computer and Internet-related criminal law, there is a debate as to whether the distribution of responsibilities set out in Section 7 *et seq*. of the Telemedia Act are also of relevance to criminal law.[50] These provisions are mainly about responsibility for the content of transmitted or temporarily stored information, not about responsibility for damage deriving from malfunctions of the network or its management. For this reason, the general principles[51] apply, and particularly the principle of trust, whereby each actor can generally rely on third parties fulfilling their duties of care if there are no indications that this is not the case.[52] The remote-controlling user can thus generally place his trust in the functioning of networks; but at the same time, every reasonable user must not only foresee certain delays in transmission, but also the need for occasional intervention by technical advisers. So the concept of negligence in criminal law cannot resolve all the problems, and cannot always unambiguously attribute responsibility.

3. Problems with Evidence

There are also significant problems with evidence when it comes to remotely controlling machines via the Internet. As is the case with other cybercrimes, it is often difficult to establish who was controlling the robot, or who may be responsible for preventing such remote control from taking place. We shall be learning a lot about this, e.g. the possibilities to resolve these problems by retaining data[53] or other virtual surveillance measures,[54]

50 Regarding the rules on responsibility in the Telemedia Act and their relationship to general criminal law *Sieber*, in: Hoeren/Sieber/Holznagel (ed.), Handbuch Multimedia-Recht, 39. Ergänzungslieferung 2014, Part 19.1 para. 14 *et seq*. with further citations.

51 Regarding the general requirements to be met by a duty of care cf. e.g. *Hardtung*, in: Joecks/Miebach (footnote 43), section 222 para. 11 *et seq*.

52 *Hardtung* (footnote 51), section 222 para. 27 with further citations.

53 Cf. in particular the panel discussion of the Crime Sciences Institute and the Institute for Legal Informatics of 20 November 2014 on "opportunities and risks of data retention".

54 Cf. e.g. the presentation by Prof. Dr Jan Dirk Roggenkamp of 4 December 2014 on "Investigations in social networks".

during the course of this series of events; I therefore shall not – and cannot due to a lack of time – pre-empt the analysis or the proposed solutions.

II. Autonomous Machines

The attribution of responsibility is even more difficult in the case of autonomous machines. First of all, it should be pointed out that, at least in the case of autonomous machines used for legal purposes and with legal objectives – electronic agents, but also robots acting in reality – criminal law must generally be applied with caution.

As with the remotely controlled machines, in the case of some detailed issues it is necessary to discuss whether they can be answered by adapting the interpretation or whether this stretches the wording too far. By way of example, let us look at the interpretation of Section 32 of the Criminal Code. Whether, for example, the targeted shot fired by an autonomous weapons system which obeys the general command of a person should be categorised as a danger deriving from an object and not as an attack can be doubted – the targeted approach taken by such machines is, at least in terms of the threat situation, rather comparable with an attack by a human being.[55] An argument against the applicability of Section 32 of the Criminal Code could however be that defence against a machine is not simultaneously defence of the legal order, since the machine itself at least is not currently recognised as a subject of that order. Also in terms of the wording, it seems unlikely that existing criminal law will truly be able to resolve this problem. Such detailed questions will arise in many areas; I will just mention the previously discussed examples of categorising certain conduct by machines as a criminal offence, the question of whether a machine can also commit a criminal trespass (see above), whether it can "stalk" someone, or whether just sitting in a self-driving vehicle equates to "driving" a vehicle.

These are just a few examples of the possible detailed questions which could arise in criminal law due to the use of autonomous machines. In

---

55 Since autonomous weapons systems also obey general commands issued by a person, the situation is comparable to attacks by animals which, at least when the animal is deliberately used as a means of attack, can represent a self-defence situation pursuant to Section 32 of the Criminal Code, *Perron* (footnote 31), Section 32 para. 3.

general, unlike in the case of remote control, it is probably the case that it is not possible to impose a punishment for a deliberate crime *de lege lata* even if, due to the objectives assigned to it, the machine acts in a manner that would amount to a crime if it were done by human being. Current criminal law is simply unable to cover these machines.[56] Only in the case of the deployment of an automatic weapons system could one discuss the responsibility which exists under international criminal law.[57]

Ultimately, current criminal law offers two main alternatives for the violations of protected interests caused by autonomous machines: an adaptation of the concept of negligence and a general ban on the use of autonomous and networked autonomous machines, at least in certain areas of life.

The application of current law to autonomous machines also creates substantial difficulties with regard to the punishability of negligence. For example, no set of social rules currently exists for the emerging technology which could provide guidance to programmers, manufacturers and users – for this reason, it is extremely difficult to draw the line between socially reasonable and reprehensible conduct.[58] Proving who contributed exactly what to the occurrence of damage caused by the machine is impossible in most cases, because not only do numerous people work together on manufacture and deployment, but also the machines act in a highly complex and ultimately unpredictable way.[59] The unpredictability is intended: the machines of the future are supposed to be equipped with the ability to adapt and to learn.[60] They will also, as we are considering here, be networked with one another, and this will exacerbate the unpredictability and mean that errors propagate and only the cumulative action will then cause the violation of the protected interest. It is also feasible that the networking not only will be particularly dangerous, but is also undesired by society because of the possible violation of other protected interests (privacy).

56 *Joerden*, Strafrechtliche Perspektiven der Robotik, in: Hilgendorf/Günther (footnote 10), p. 203 *et seq*.

57 Autonomous weapons systems cannot comply with the rules of international humanitarian law because they are unable to distinguish appropriately between soldiers and civilians.

58 Cf. in this regard *Joerden* (footnote 56), p. 207 *et seq*.

59 *Beck* (footnote 4), pp. 102 f.

60 Cf. Neuhäuser (footnote 7), p. 270.

If, however, unpredictability is part of the design concept, not every "error" by the robot will be caused by misconduct by the participants:[61] it may be the case that machines simply decide differently from people. So it is neither possible simply to demand that the construction of unpredictable machines *per se* be prohibited, nor – if the machines meet with social acceptance – can the damage they cause be attributed in the same way as when it is caused by a person. It is equally impossible to use current criminal law to construct something along the lines of the English "strict liability",[62] because this would contradict the principle of culpability. Certainly, it is not possible to derive liability from the purely abstract predictability of some sort of danger when autonomous machines are deployed or networked. Also, it is implausible that the beneficiaries of particularly dangerous technology will be able to fully escape liability by referring to the socially acceptable nature of their conduct. A fresh discussion is needed of the question of how specifically predictable the occurrence or the violation of the protected interest needs to be in order to be liable to punishment – and to some extent some fine-tuning has already started.[63]

Further, excessively risky deployment of such machines which only benefits certain groups could be subject to a general prohibition; new legislation would need to be put in place for this.

Overall, it can be stated here that current criminal law can only provide a very limited response to the conduct of autonomous machines – so there is clearly a need for a certain adjustment and fine-tuning.

E. Limitations to Criminal Law

However, there are also limits to this adjustment, and even to the adoption of new rules. In fact, criminal law as a whole is unable to cope with new technologies: criminal law is oriented to past acts of an individual offender; it must be possible to precisely describe these acts; due to this principle of certainty, it is difficult to adapt the law to new developments; it is na-

61 *Beck* (footnote 18), pp. 14 f.

62 Further detail in this regard in *Radtke*, in: Joecks/Miebach (footnote 24), section 14 para. 138.

63 *Joerden* (footnote 56), p. 207 *et seq.*; *Beck* (footnote 4), p. 108 *et seq.*

tionally oriented. In contrast, new technologies are forward-looking, develop quickly, and in some cases are unpredictable and international.[64]

An important boundary here is the need for individual accountability. This contrasts with the fact that the networking of human beings with machines, or the deliberate transfer of decision-making to individual or networked machines tends to result in a diffusion of responsibility. The section which is covered by *ex post* criminal law (the human action, which only accounts for a small proportion of the damage) fails to cope with the reality of such situations. If the responsibility is spread among several people, it is feasible that none of them acts reprehensibly within the meaning of criminal law, that no action creates a "legally relevant risk",[65] or that the outcome is not the realisation of the risk of a single act. So criminal law reaches its limits even at this point, and in particular, it cannot, as mentioned, provide the grounds for liability not related to culpability.

The punishment of a collective or even of the machine itself, as is in principle possible from a constructivist viewpoint,[66] entails certain dangers. After all, existing criminal law traditionally not only fulfils security-related functions, but also aims to guarantee the freedom of the (potential) offender.[67] This goal can only be pursued if criminal law is not focused primarily on safeguarding society, but focuses on the offender and takes his dignity and autonomy seriously by punishing him for culpable, clearly reprehensible, acts.

I. With Human Involvement

In the processing of deliberate crimes which are committed by a remotely controlled machine, criminal law comes up against similar boundaries as are encountered in cybercrime in general: a lack of social control,

---

64 Cf. *Beck*, Dealing with the diffusion of legal responsibility: the case of robotics, in: Battaglia/Mukerji/Nida-Rümelin (ed.), Rethinking Responsibility in Science and Technology, 2014, p. 167.

65 Regarding the existence of a legally relevant danger as a precondition for accountability *Eisele*, in: Schönke/Schröder (footnote 31), before section 13 *et seq*. para. 93.

66 *Hilgendorf*, Können Roboter schuldhaft handeln?, in: Beck (footnote 18), p. 127 *et seq*.

67 Regarding the various purposes of criminal law *Hörnle* (footnote 26), p. 15 *et seq*.

anonymity, problems with evidence, and – not least due to the national nature of criminal law – inadequate protection for victims.

It is also virtually impossible in the case of remote control to precisely describe the boundary beyond which conduct becomes too risky to be deemed socially adequate. But new prohibitions would have to be measured against the principle of certainty. At the same time, it is also not possible simply to refer to certain standards of a certain segment of the population, since such standards have yet to be formed for this area of life. Yet liability for negligence must not lose all shape and must not create a culpability independent of standards which no longer refers to a violation of social and moral rules but punishes the hazardousness of the act itself, irrespective of motivation, benefits or social acceptance. Criminal law reaches its limits here: with a view to the principle of liability, the last-resort principle and the principle of legality, it must not move too far away from unambiguous errors in the exercise of remote control.

II. Machines

The accountability problems can be seen even more clearly in the assessment of the use of autonomous machines in terms of criminal law: once the machine has been set in motion with a specific objective, its actions can only be regarded as a deliberate diffusion of responsibility. If it is not exceptionally possible to prove an unambiguous error by a participant, there can be no culpability here without interpreting this transfer of responsibility *ad absurdum*. The problems of extending current criminal law to the collective or the machine have already been discussed.

It will be difficult to justify prohibitions on the use of autonomous machines in certain areas of life. Such prohibitions represent a massive restriction on freedom to engage in research and on the general freedom of the participants to act. Also, they deprive society of the advantages of a machine-taken decision. And criminal law is the last resort of the state, and may only be applied in the case of clearly reprehensible acts. In particular, the threat of a "decline in standards in society" is not sufficient grounds for a penal law. If anything, only a ban on armed autonomous machines can be justified.

F. Changes

So what can criminal law do to provide legislative support for these advances in technology? What networking between people and machines or between autonomous machines can it prohibit, and what gaps in responsibility can it close?

I. With Human Involvement

A comparison with the cybercrime debate suggests that society will not accept a full abandonment of criminal law in this area of life. Despite the problems in the field of cybercrime, people repeatedly call for criminal law to be applied there.[68] Quite simply, certain violations of protected interests are so serious and threaten society to such an extent that it is not possible to justify a withdrawal of criminal law.

Such a massive threat certainly exists in the case of deliberate remote control of machines in order to violate protected interests. For this reason, such acts should, if they can be described sufficiently precisely, be covered by criminal law *de lege ferenda*. For example, thought could be given to an extension of Section 123 of the Criminal Code to cover entry by means of a machine or the unlawful operation of a machine in another's dwelling. An explicit inclusion in the sections on stalking and on road traffic offences also seems feasible. With regard to the practical implementation of criminal prosecution and the raising of the awareness of the Internet users that their conduct is also subject to social and legal control, reference is made to the following presentations.[69]

A certain adaptation of the negligence regime to include errors in the exercise of remote control is also feasible. For example, the dependence of the user on the functioning of the network is so great, and the user will

---

68 Matthias even advocates the introduction of a separate system of criminal law for machines. *Matthias* (footnote 16), p. 249; a sceptical view *Hilgendorf* (footnote 66), p. 130.

69 Cf. in particular the lecture by prosecutor Andreas May of 11 December 2014 on "Problems of criminal procedure in prosecuting computer crime"; cf. also the lectures by Prof. Dr Brian Valerius of 13 November 2014 on "Borderless exchange of information and borderless culpability? How far does national jurisdiction extend on the Internet?" and by Prof. Dr Jan Dirk Roggenkamp of 4 December 2014 on "Investigations in social networks".

undoubtedly be so aware of this, that one could at least discuss a reversal of the principle of trust;[70] this would mean that the user himself would be responsible for either only deploying machines which do not violate any protected interests even if the network fails or if technical assistants are involved, or to inform other participants fully about such risks and to obtain their corresponding agreement – or in a case of doubt, given a substantial danger to significant third-party protected interests, he would have to refrain from deploying the machine. Apart from this, rules outside criminal law, such as a revision of road traffic law, data protection law, civil-law liability for causing danger, are probably more important in terms of controlling the remote control of machines.[71]

II. Autonomous Machines

It is even less possible to cover the use of autonomous (networked) machines in terms of criminal law. As long as society accepts such machines as being useful, it has to accept the related transfer of responsibility to these machines, and may not pass this burden on to individuals. The liability of persons who manufacture, sell and use autonomous machines must, however, in those areas in which the machines are in principle allowed, necessarily be less great than it would be had they acted themselves. This is particularly true of criminal liability targeted at individuals.

This does not mean that there is no liability at all for negligence; the more dangerous the deployment of autonomous machines is in a certain area of life, the more it corresponds to the benefit of those deploying the machine, the more uncontrollable the machine becomes for those deploying it and for third parties, the more, despite the lack of predictability of specific errors and damage, liability for negligence remains plausible. In particular, it is presumably obvious to every user that wrong decisions can be made by machines which are armed and which thus solely serve the purpose of injuring or even killing people sooner or later. If such machines are used solely for private purposes, it therefore seems likely that liability for negligence relating to wrong decisions will have to be borne, even in terms of the existing negligence regime. This is particularly the case when

70 *Duttge*, in: Joecks/Miebach (footnote 24), section 15 para. 140 *et seq*.

71 Regarding the extension of liability for endangering under civil law *Bräutigam/Klindt* (footnote 11), p. 1138 *et seq*.

these machines are networked with other machines or the general Internet, since it is again manifest that there is a possibility of third-party access, erroneous information or harmful cumulative effects. However, an adaptation of the negligence regime should be rejected for machines that are beneficial to society as a whole and only by chance do damage to the protected interests of third parties. If, for the benefit of society as a whole, an automation or even a networking is necessary, a certain individual participant should not subsequently be blamed for this.

The fact that the liability of a person cannot simply be replaced by the punishment of the machine has already been discussed.[72] It might be feasible to create special criminal law for such entities, as already exists for collectives in some countries.[73] Here, there could be a sort of intermediate stage between administrative offences law and criminal law, which compared with the former would have greater communicative force,[74] and compared with the latter would "cope" with a certain renunciation of the principles of culpability and legality.[75] However, it will only be possible to assess whether this new system is necessary over the course of time. These developments can be awaited; there is no need to be afraid of realities which have already been created, since rules can already be adopted in other fields of law and these can give better support to a generally accepted technological development than criminal law.

In a few areas, however, either the deployment of autonomous machines *per se* or their networked activity could be prohibited under crimi-

---

72 Cf. the comments under C. II.

73 Genuine collective culpability has traditionally existed in the United Kingdom and Ireland. Collective culpability has also been introduced in the Netherlands, Portugal, Norway, Iceland, France, Finland, Denmark, Slovenia, Belgium, Malta, Switzerland, Spain, Liechtenstein and the Czech Republic, although the provisions reveal some substantial conceptual differences. In Sweden, corporate sanctions are a *sui generis* consequence of culpability. In Italy and Austria, special laws have been passed on collective responsibility. Eastern European countries (Estonia, Lithuania, Latvia, Hungary) have since followed suit. Cf. in this regard *Heine/Weißer*, in: Schönke/Schröder (footnote 31), before section 25 *et seq*. para. 124 with further citations.

74 Regarding the communicative force of criminal law *Hörnle* (footnote 26), p. 7 *et seq*.

75 Regarding the culpability principle *Radtke* (footnote 23), before section 38 *et seq*. para. 14; regarding the principle of legality *Schmitz*, in: Joecks/Miebach (footnote 24), section 1 para. 1 *et seq*.

nal law[76] – taking account of the certainty requirement. This is always the case when and where society does not wish to accept the related violations of protected interests. For example, it would be quite reasonable to prohibit autonomous machines in general, i.e. in times both of peace and of war, from deciding on life and death, or not to allow them to bring up small children or to serve as personal carers for sick or elderly people. Also, the networking of machines in the medical sphere can be restricted, or the networking can be restricted where this means the disclosure of personal data; this is because the danger may lie not in the automation, but in the networking.

Prohibitions under criminal law should, however, be restricted to those areas where there is a consensus in society that no automation should be allowed. Also, the prohibited action must be described precisely. This will be difficult in the case of such far-reaching rules, but is necessary in view of the requirement of certainty.

G. Conclusions

The development of increasingly autonomous machines will be one of the most important technological revolutions of the coming years. The law will scarcely be able to prevent this development from taking place – and a comprehensive ban on this technology would probably be undesirable in any case. But there is no denying that the use of such machines – particularly when they are networked with one another or with the Internet – entails considerable risks. The individual actions of one of these machines, whether it is remotely controlled by the user via the Internet or whether it is autonomously pursuing specific objectives, are no longer truly predictable and controllable. Not just autonomous weapons systems, but also electronic software agents or remotely controlled service robots have the potential to violate various protected interests of the users and third parties.

Despite these dangers, criminal law is rather unhelpful as a means of regulating this development. It can undoubtedly prohibit those actions

76 For example, Human Rights Watch and the International Human Rights Clinic call for an international agreement to ban "killer robots", https://netzpolitik.org/2012/kriegsroboter-und-drohnen-human-rights-watch-fordert-verbot-von-killer-robots/ (18 May 2015).

which are regarded by society as a whole as being too risky, offering too little benefit, or being "hostile to human beings", as long as these actions can be described sufficiently precisely. But beyond this, the increasingly frequent division of responsibility between people and machines means in most cases that an individual cannot be subject to criminal prosecution, since that person's conduct on its own is not sufficiently reprehensible. In general, in the assessment of Google Cars, software agents and autonomous weapons systems – and all similar machines – the problems of cybercrime and those of making machines autonomous increase exponentially, and this creates significant problems for the science and application of criminal law.

However, this should not cause one to lapse into pessimism about the future of society or even pessimism about criminal law. Rather, the science of criminal law continues to be called on to identify those areas of life in which the communicative force of criminal law can actually contribute to the stabilisation of society. In the context of networked partially autonomous and autonomous machines, this might mean, for example, completely banning the use of autonomous weapons systems or private weapons that can be controlled via the Internet. Apart from this, a threat of liability for negligence might result in a careful design of the machines – and so early cooperation with the institutes which design the relevant non-statutory standards is required. The networks via which the machines communicate with people or with one another must be adequately safeguarded. Here, the security of personal data should have absolute priority. The possibility of physical harm being caused to third parties must be restricted as far as is possible by appropriately soft materials and good sensors. In case of doubt, autonomous machines should only act on an auxiliary basis, without taking social or even moral decisions themselves. However, the latter boundary in particular cannot be guaranteed by criminal law, but only by the careful action of the manufacturers, consumers, and society itself.

Data Law in Business and Industry – A First Glance at a Problematic Relationship

*Berthold Haustein,**
University of Würzburg

A. Introduction[1]

It is common knowledge that data protection and thus also data protection law are becoming more and more important. With the increasing networking of everything with everything else (in the words of the German Federal Constitutional Court: "under the conditions of modern communication"),[2] there exists an omnipresent, ubiquitous exchange of data. As a component of human life, as an organizational form of human knowledge or as a commercial asset, data is the subject of communication and commercial exchange. But it is also the subject of law, which is encountered particularly in the context of two high-profile topoi. On the one hand, the protection of personal data on the basis of the fundamental right to informational self-determination ("data protection *is* the protection of fundamental rights "),[3] has been stylized as an "exceptional" foundation of social development in liberal democratic constitutional states under the rule of law. On the other hand, there is an ongoing discussion in legal science

---

* Berthold Haustein was, until February 2017, an assistant professor at the "Forschungsstelle RobotRecht" which is attached to the chair of Prof. Dr. Dr. Eric Hilgendorf at the Julius Maximilians University in Würzburg. He is now completing his practical legal preparatory service at the Higher Regional Court of Dresden. His research focuses both on creating absolute rights in data and on data protection law challenges for industry and commerce.

1 A shorter version of this article was published in Blanke, Herm.-J. & Grusser, G. (eds..), *Industrie und Wirtschaft 4.0, Politische, rechtliche sowie wirtschaftliche Herausforderungen, Tagungsberichte des Staatswissenschaftlichen Forums Erfurt e.V.*, Gera 2016, pp. 44.-46.

2 BVerfG *Neue Juristische Wochenschrift* 1984, p. 419 (419) (= BVerfGE 61, 1).

3 Cf. http://www.bfdi.bund.de/Shared Docs/Publikationen/Faltblaetter/Datenschutz-ist.pdf?__blob=publicationFile (last accessed 19.2.2016).

about the concept of "data property",[4] meaning a comprehensive allocation of property rights in data to legal subjects which is consistent with the respective general system of property law. Time and time again the latter is described as particularly important in the context of the development of autonomous technology, particularly in business and industry. It is thus closely associated with another buzzword, namely, *Industrie 4.0*,[5] which refers to the various developments in modern industrial production technology and closely related services. Thus modern industry and commerce see themselves placed into the middle of a dispute, which is sometimes conducted quite apodictically. They seek to create new or at least generally overhauled business concepts and products compliant to a partly blurred framework of data law and data protection law.

At the same time, both the law and the debate in legal science are finding it difficult to state in a tolerably abstract way, what *data* are. The criminal law can point to the legal definition or at least to the outlines of a legal definition in § 202a Sec. 2 of the German Criminal Code (*Strafgesetzbuch* – StGB). The norm applies only to data that are stored or transmitted electronically or magnetically and are not directly perceptible (to human senses). This wording, however, does not apply to all aspects of computer criminal law. For example, it does not apply (at least not directly) to the provisions contained in § 269 StGB, which make it a criminal offence to forge "data intended to provide proof". However, a certain amount of agreement can be discerned to the effect that – at least in the area of criminal law – data is defined as information existing in some sort of code.[6]

§ 3 Sec. 1 of the German Federal Data Protection Law – (*Bundesdatenschutzgesetz* – BDSG) sees this differently, but only at first glance. According to this provision, data are "all individual details", i.e. "any information".[7] However, § 1 Sec. 2 No. 3 BDSG stipulates that the law applies to private and non-governmental bodies (among others) only for data processing operations with data processing systems. This assumes the coding

4 Cf. on this subject, M. Dorner, "Big Data und ‚Dateneigentum'", in *Computer und Recht* 2014, p. 618 *et seq*.

5 See for example, H. Zech, "'Industrie 4.0' – Rechtsrahmen für eine Datenwirtschaft im Digitalen Binnenmarkt" in *Gewerblicher Rechtschutz und Urheberrecht* 2015, p. 1151.

6 T. Fischer: *Strafgesetzbuch mit Nebengesetzen*, Munich 2016, § 268, paragraph 6.

7 U. Dammann, "§ 3 – Weitere Begriffsbestimmungen" in S. Simitis (ed.), *Bundesdatenschutzgesetz*, Baden-Baden, 8th ed. 2014, paragraph 3.

of data. But the most important connecting factor remains the reference to persons under § 3 Sec. 1 BDSG. The data *vulgo* the information *vulgo* the individual details must refer to a specific or determinable natural person, in order to be relevant for data protection law. Therefore, the decisive criterion under the BDSG is a substantive one rather than a technical one, as was discussed above in reference to StGB.[8] This "mechanism" is also used by intellectual property law. For example, copyright protects a piece of work. One could say that what is protected, in the widest sense, is the content of the work. The existence of immaterial content is a condition for a work to come under the scope of protection of German Copyright Law (*Urhebergesetz* – UrhG). See, for example, § 2 UrhG.

The civil law in its narrow sense – namely the German Civil Code (*Bürgerliches Gesetzbuch* – BGB) – does not have a concept of data. Since "day 1" of digitization, there has been an argument ongoing, for example, whether the data stored on a data carrier – such as a software program on a CD, to choose a somewhat old-fashioned example – are part of what is owned and at the disposal of the owner of the data carrier under § 903 BGB.[9] Some argue that (even) stored data are in and by itself not "*things*" (roughly: *res corporales*) under § 90 BGB and therefore, strictly speaking, cannot be "*owned*". Despite this uncertainty in owning data, it is, however, reasonably certain nowadays that data may be *sold* independently of a data carrier, namely under § 453 Sec. 1 BGB. In this sense they are classified as "other objects"[10] in German law.

It was against this background that political and legal science discussions, as mentioned above, began on how a comprehensive system of rights in data could be created, which would be comparable to the system the civil law applies to chattels. Proposals and demands have been as numerous as they are varied. It has to be assumed that sooner or later a property right in the form of "data ownership" will be created which as a systematic construct will be consistent with general German civil law doctrines, although it cannot be predicted exactly how it will deal with the principles of informational self-determination and data protection. How-

8 Zech refers to "semantic" in H. Zech, *Informationen als Schutzgegenstand*, Tübingen, 2012, p. 353.

9 M. König, "Die Qualifizierung von Computerprogrammen als Sachen i.S. des § 90 BGB" in *Neue Juristische Wochenschrift* 1989, p. 2604.

10 See T. Hoeren, "Dateneigentum, Versuch einer Anwendung von § 303a StGB im Zivilrecht" in *MultiMedia und Recht* 2013, p. 486 (489).

ever, it should be noted that developments in the law of such disruptive nature usually have a certain tendency to come about rather slowly. For this reason, they find themselves as being regularly exposed to criticism in society, especially by private-sector actors. In point of fact, the divergence between legal and technical development is certainly one of the challenges that business have to face in the age of modern communications, not exclusively but especially in regard to data protection law.

It is true, for example, that the Federal Constitutional Court's judgment on the 1983 census, the theoretical superstructure of personal data protection, is over 30 years old, and has hardly adapted since then. It is also true that the BDSG was significantly amended for the last time in 2001,[11] and that the legal provisions for dealing with human images dates all the way back to 1907. Similarly, private commercial exchange of data in German civil law takes place under the framework of provisions in the original Civil Code which came into effect in 1900. It is equally true, however, that historically some of the developments of the past which were proclaimed as disruptive were very short-lived (and vice versa). It is also part of the ordering function of law to provide continuity and consistency in the context of dynamic social and economic change. Therefore, for the time being, the providers of modern products and networked services will have little choice but to deal with the legal situation as it is today.

Several points indicating why this may lead to problems are raised below, using examples, but without any aspirations to being even close to a complete treatment.

B. Data Law in Business and Industry

Issues about rights in data can be abstracted to a certain extent and classified according to the actors involved. Private commercial data exchange plays a role in three more or less bipolar relationships: between firms and customers, between firms and employees, and between firms and other

11 J. Taeger, *Datenschutzrecht, Einführung*, Frankfurt a.M., 2014, p. 5. This, of course, will change significantly with the coming into force of the EU's General Data Protection regulation as of May 2018 (Regulation (EU) 2016/679 of The European Parliament and of the Council of 27 April 2016 on the protection of natural persons with regard to the processing of personal data and on the free movement of such data, and repealing Directive 95/46/EC).

firms. Simplified slightly, it can be said that the first relationship comprises mainly "traditional questions" of personal data protection. In contrast, questions of employee data protection, in the second relationship, raise many different and cross-cutting legal issues. Lastly, the third relationship has, above all, civil law and competition law implications. It raises a potpourri of systems issues and questions on the lawful handling of information and data, which could hardly be more complex or more multi-faceted. In the traditional environment of a firm that produces products for the consumer market, there will be extensive rules on how the personnel department is supposed to deal with employee data. The data protection officer will keep an eye on the storage, processing and deletion of billing data of its customers and the legal department takes care of the registration of brands and designs of new products. The structure of the treatment which follows will also be aligned to these three functional areas. However, it has to be said that one of the characteristics of data-based business is that the lines of separation between functional areas are no longer as clear as they once were. Data are not only available at every interface in the firm, they are also needed and used everywhere, and thus they stimulate new questions: How should the original design rights of a customer on a largely individualized product be dealt with? How to achieve a balance between the principles of data minimization and the harm caused by constant surveillance of workers to an extent where there is no privacy at the workplace left at all, on the one hand, and the requirements of sensor-based production equipment and systems on the factory floor, on the other? How can informational self-determination be reconciled with the domestic nature of data protection laws in an international production and logistics chain?

C. Data Protection of Personal Data, Customer Data and Highly Individualized Products

The term "data protection" (*Datenschutz*) is associated in Germany first and foremost with the protection of individual privacy rights, particularized as the basic right of informational self-determination against the unauthorized processing of personal data by third parties. In its current form, data protection in Germany is regulated primarily by the BDSG.[12] With the coming into force in May 2018 of the EU's General Data Protection Regulation (GDPR),[13] regulatory provisions in the area of "data law 2018" will finally be fully harmonized across the EU.

If one looks at "*Economy 4.0*", several developments become apparent which may present special challenges in the area of personal data protection. In the service sector, on the one hand, there is increasing collection of personal data, which can then be monetized through targeted advertising. Of course, this is not a new development, but has been part of the business model of firms on the internet since its inception. However, the phenomenon is becoming qualitatively different due to the increasingly oligopolistic market structure of the internet as well as the expansion of linked or at least linkable data collection via portable end-user devices such as mobile phones or embedded chips (so-called "wearables"). In addition, big-data analysis will provide a treasure trove of information not obtainable by other means.[14] It is only at first glance that this remains largely in the area of service provision. As the phenomenon of the "Internet of Things" develops, the distinction will become increasingly meaningless, as the boundary between service and production continues to blur further ("everything as a service").

In the context of industrial production, the increasing individualization of products has also become relevant recently. This is being discussed using the keyword *lot size 1 production*. What is meant is the goal of being able to customize products for each customer within the industrial production process without any loss in time or efficiency compared to serial production. A particularly instructive example of this is the individualized production of sneakers as ordered by customers, which is already in the market. The data required for this are sent to the production site via the internet when the customer fills out an online order form. This is certainly not a major new data protection law challenge, but the quantitative change in data throughput in the company might make it necessary to rethink some data protection law provisions. If lot size 1 is directed at final consumers, more personal data will need to be processed. And it will be pro-

12 Implementation of Directive 96/46/EU. Currently, the EU is seeking full harmonization through the General Data Protection Regulation (GDPR).

13 Cf. fn. 11.

14 Cf. also C. Peschel & S. Rockstroh, "Big Data in der Industrie, Chancen und Risiken neuer datenbasierter Dienst" in *MultiMedia und Recht* 2014, p. 571 *et seq.*

cessed in all departments of the company and not only in the shipping and accounting departments.

This is hard to reconcile with the fundamental principle of the German and European data protection law systems. Under § 4 Sec. 1 BDSG and Art. 6 Sec. 1 GDPR the legality of each instance of the collection, transmission, storage or processing of each personal datum must be checked separately in regard to its lawfulness. Every one of them, in principle, is considered unlawful unless either a legal norm permits it, or the person whose personal data are being processed has consented to the concrete processing for a specific and determined purpose. This system has been frequently (and not always unjustly) criticized as inflexible and ill-suited to modern data processing.[15] It is based on a paradigm established by the Federal Constitutional Court in the early 1980s that there are "no unimportant data".[16] Therefore the same set of rules applies to all personal data regardless of their sensitivity in regard to individual privacy.

In practice, this condition is mostly met by gaining the consent of the data subject in accordance with § 4 Sec. 1 BDSG read in conjunction with § 13 Telemedia Act (*Telemediengesetz* – TMG).[17] Anyone who uses the internet at all will click their way almost daily through the corresponding data protection statements which have to explain in generally understandable language the nature, scope and purpose of the each specific act of data processing. In practice, they are true "monstrosities". Whereas the data protection terms of the German Federal Ministry of Economics and Technology's website has "only" 651 words, the American dating service "Tinder" expects users to cope with 3,076 words before they can meet the love of their live in a data protection compliant way. The German website of "Ebay" informs users about its implementation of data protection law in 5,334 words. To give readers a yardstick for comparison: the one-page article "Questionable property of data" by Daniel Zimmer in the business section of the *Frankfurter Allgemeine Zeitung* of 18 November 2016 came to just 2,093 words. Incidentally, "Ebay's" data protection statement is

15 For a sophisticated treatment cf. J. Masing, "Herausforderungen des Datenschutzes" in *Neue Juristische Wochenschrift* 2012, p. 2305 (2307).

16 BVerfGE *Neue Juristische Wochenschrift* 1984, p. 419 (422).

17 Legal provisions exist to regulate this, for example § 28 (1) BDSG, but they often do not go as far as the providers of data-driven services would like, cf. for CRM, T. Söbbing, "Customer-Relationship-Management (CRM) – Rechtliche Möglichkeiten beim CRM" in *Recht der Datenverarbeitung* 2016, p. 120.

also longer than the article you are currently reading, at least the original German version of which amounted to only 3,637 words.[18]

The problem is well recognized:[19] this sort of consent does hardly contribute to real informational self-determination. It is, nevertheless, very helpful to the one or the other business that doesn't have customer data protection at the top of their corporate values. Proposals on how improvements in the reality of data protection law can be achieved range from the introduction of "data protection traffic lights", all the way to anchoring the statements in the law of the (also rarely read) general terms and conditions contained in consumer contracts. In addition, GDPR, also introduces technical solutions. The concepts of "privacy by design" and of "privacy by default" are receiving equal attention. Under Art. 25 Sec. 2 GDPR, service providers will have to choose the most data protection-friendly default setting.[20] The additional requirement of interoperability[21] (Art. 20 GDPR) means than in the future, data subjects will have better access to stored data. New data formats are also being considered, for example, formats that self-delete after a certain amount of time has passed or a certain condition is fulfilled, thus contributing to more technical data protection.

D. Employee Data and Networked Production Methods

Employee data protection issues have been the focus of discussion in legal science for quite some time. The relationship between employer and employee is special in many respects, first and foremost this regards the actual distribution of power between them. Although an employment contract

---

18 The words of these protection statements have been carefully counted on Dezember 27th 2016 but may have changed since them. The author chooses to spare the reader with hyperlinks to their origins. They can be easily found on the respective websites.

19 S. Simitis, "§ 4a – Einwilligung" in: S. Simitis (ed.), *Bundesdatenschutzgesetz*, Baden-Baden, 8th ed. 2014, paragraph 4 *et seq.*

20 C. Barlag, "§ 3 – Allgemeine Regel der Datenschutzgrundverordnung" in A. Rossnagel (ed.), *Europäische Datenschutz-Grundverordnung, Vorrang des Unionsrechts – Anwendbarkeit des nationalen Rechts*, Baden-Baden, 2016, paragraph 228 *et seq*.

21 T. Jülicher, C. Röttgen & M. v. Schönfeld, "Das Recht auf Datenübertragbarkeit, Ein datenschutzrechtliches Novum" in *Zeitschrift für Datenschutz* 2016, p. 358 (360). They also see an evolution in the right to informational self-determination.

is a normal civil law contract, it can hardly be said that it meets the ideal of equality between contract parties. Consequently, there are special rules which make employee data protection matters a particularly interesting area.

This is all the more true if one considers how modern industrial production is conducted today. One of the most distinguishing developments in this context is the increasing capacity of production equipment to collect and process data about employees.[22] The worker stands in the midst of a soon to be fully networked workshop with autonomously moving logistics robots, intelligent shelves, a uniform control system and so on.

In this regard one particular norm of the Works Constitutions Act (*betriebsverfassungsrechtliche Normen*) governing the co-determination and thus the protection of workers and their data from technical monitoring devices, § 87 Sec. 1 No. 6 of the Works Constitution Act (*Betriebsverfassungsgesetz* – BetrVG), is gaining momentum.[23] This norm establishes a workers' co-determination right in the event of the introduction of technologies at the workplace in order to monitor behaviour or performance of workers. Following a judgment of the Federal Labour Court in the 1970s,[24] the norm is interpreted in its broadest possible way. Technologies with the mere capacity to monitor workers fall under its scope regardless of whether they are used to this effect. In some cases even a situation where employees feel monitored by a certain device, suffices to trigger the right of co-determination.[25] Whether this interpretation was appropriate in the context it was given is a moot point. In light of *Industrie 4.0*, however, it is problematic: networked production resources will increasingly collect data about their environment, and in doing so they will necessarily collect

---

22 Cf. K. Hofmann, "Smart Factory – Arbeitnehmerdatenschutz in der Industrie 4.0, Datenschutzrechtliche Besonderheiten und Herausforderungen" in *Zeitschrift für Datenschutz* 2016, p. 12 (16). His demand, for example, is that collision avoidance systems be designed in such a way that an identification of the worker is not technically possible. This is at best not practical, and at worst (technically) not feasible.

23 On the issue of work and data protection in the smart factory, cf. K. Hofmann, "Smart Factory – Arbeitnehmerdatenschutz in der Industrie 4.0, Datenschutzrechtliche Besonderheiten und Herausforderungen" in *Zeitschrift für Datenschutz* 2016, p. 12 *et seq*.

24 BAG, AP Nr. 1 zu § 87 1972 (labour court's judgment on surveillance).

25 M. Kort, "Betriebsrat und Arbeitnehmerdatenschutz" in *Zeitschrift für Datenschutz* 2016, p. 3 (5).

data about their operators or the employees they work with.[26] In the end, then co-determination rights contained in § 87 Sec. 1 No. 6 BetrVG will extend to practically every component of (modern) production equipment, which would probably lead to the unintended consequence of a shift in power relations within the company from management to the works council.

The norm, however, is only one example of a more general issue: the increasing availability of production equipment with sensors, together with the growth of networking, will lead to a situation in workplaces of the future in which more and more data about employees will be generated. This data will not solely be a "by-product" of industrial production, it will rather be a necessary input to maintain the operation of the machines used. Data protection law does not provide a norm appropriate to this development.[27] As it is today, the closest data protection law comes to regulating sensors at workplaces is § 6b BDSG. The norm is applicable in cases where cameras are employed and, using the keyword of "video surveillance", differentiates, between publicly accessible and not publicly accessible rooms, like a factory hall. It is primarily oriented toward the detection of criminal offences, not toward co-working with robots.[28]

The use of so-called "wearables"[29] in the workplace is another instructive example of this development. Here, technology is increasingly becoming one with the human being or – for that matter – the employee.[30] What is crucially important here, as it is in the above-mentioned situations, is that an employee's consent, for example, to the use of data glasses is as a rule not recognized as valid legal consent in data protection law. As the employee is dependent on the employer, he is deemed to usually have no real alternative to consent without the risk of losing his job. Therefore,

---

26 For a different opinion, cf.: K. Hofmann, "Smart Factory – Arbeitnehmerdatenschutz in der Industrie 4.0, Datenschutzrechtliche Besonderheiten und Herausforderungen" in *Zeitschrift für Datenschutz* 2016, p. 12 (15), for whom under § 87(1) (6) BetrVG technical surveillance devices should be used in significantly fewer design variants.

27 Regarding § 32 BDSG cf. Taeger (Fn. 5) p. 112 *et seq*.

28 Cf.: F. Klein & P. Roos, "Videoüberwachung: Kostspielige Folgen für den Arbeitgeber? Aktuelle Rechtsprechung – konkrete Bemessungsmaßstäbe" in *Zeitschrift für Datenschutz* 2016, p. 65 (67.).

29 Alternatively "mobile digital devices".

30 Cf. on this subject R. Kopp & K. Sokoll, "Wearables am Arbeitsplatz, Einfallstore für Alltagsüberwachung?" in *Neue Zeitschrift für Arbeitsrecht* 2015, p. 1352.

the consent cannot be considered as given voluntarily.[31] For the company's management it is therefore necessary to look for other legal grounds to make the data processing lawful. Such grounds may for example derive from internal company agreements with works councils.[32] Here, it becomes clear again that the legal doctrine and systematics, separating strictly different instances and occasions of data processing, are diametrically opposed to the technological development, as the purpose of the latter is precisely to reconcile widely disparate goals. For example, data glasses can improve the speed of sorting tasks, prevent pilfering, increase accuracy, facilitate inventory replacement, update service plans, organize reorders, etc. However, it is also possible to create movement profiles of employees with these glasses. And this is only lawful in very exceptional cases under German data protection law, thus.[33]

Allowing for the author's modest prophetic pretensions, the issue of "employee data protection" will develop in the near future into a central question of data law and data protection law systems in Germany and Europe. On the one hand, the new technologies in industry promise increases in productivity and new business models, which can significantly contribute to Europe remaining a major technology centre and thus secure employment and economic growth. On the other hand, they call into question the participatory co-determination structures existing, for example, in the German working world by giving employers the opportunity to collect almost unlimited amounts of information on the activities, performance and behaviour of their employees. Achieving a balance that is satisfactory to these conflicting interests is a task of Gordian proportions.[34]

---

31 R. Kopp & K. Sokoll, "Wearables am Arbeitsplatz, Einfallstore für Alltagsüberwachung?" in *Neue Zeitschrift für Arbeitsrecht* 2015, p. 1352 (1354).

32 P. Scholz & K. Sokol, "§ 4 – Zulässigkeit der Datenerhebung, -verarbeitung" in S. Simitis (ed.), *Bundesdatenschutzgesetz*, Baden-Baden, 8th ed. 2014, paragraph 4.

33 R. Kopp & K. Sokoll, "Wearables am Arbeitsplatz, Einfallstore für Alltagsüberwachung?" in *Neue Zeitschrift für Arbeitsrecht* 2015, p. 1355.

34 For different approaches to achieving such a balance, cf. K. Uffmann, "Digitalisierung der Arbeitswelt, Wie gestalten wir die notwendigen Veränderungen?" in *Neue Zeitschrift für Arbeitsrecht* 2016, p. 977. For a discussion of the changes which will follow from the introduction of GDPR see also T. Wybitul & S. Pötters, *Der neue Datenschutz am Arbeitsplatz, in: Recht der Datenverarbeitung* 2016, p. 10.

E. Data Protection without Informational Self-Determination

Where data are exchanged or processed between firms, "traditional" data protection law can be applicable, but it need not be. Where no personal data are processed, that is, no data relating to a specific or determinable person (§ 3 Sec. 1 BDSG or Art. 4 Sec. 1 DsGVO) the German data protection law does not apply. These data then exist in a largely unregulated area. If an agency sells address data to a firm for advertising purposes, BDSG provides more or less clear rules for when this is permitted and what the buyer may do with the data. If, on the other hand, non-personal data, such as the degree of wear or life expectancy of machinery are at issue, the entirety of German data protection law does not apply, despite the fact that these data may be of significant value.[35]

This area of non-regulation exists in legal regard mainly due to the fact that, unlike, for example, "chattels" which are someone's property, there is no comprehensive legal system or rule for determining the ownership of data. This lacuna is closed within certain limits in the case of personal data by data protection law, in particular by BDSG, but in the case of non-personal data, the lacuna is still wide open. Within the framework of the criminal law, which is mainly laid down in §§ 202a-c, § 303a StGB and in § 17 of the Act Against Unfair Competition (*Gesetz gegen unlauteren Wettbewerb* – UWG),[36] it is suggested that the handling of data be regulated through contracts, creating obligations between the parties. German legal science is already engaged in an intense debate on how to change this.[37] The development of a "proper" data law would play a pivotal role in changing how the law regards and handles (non-personal as well as personal) data, namely as a commercial asset. This will be even more the case as more data are generated by *Industrie 4.0* production equipment and

---

35 There exist of course exceptional situations in which both personal and non-personal data are purchased, such as in the case of mergers or takeovers, cf. I. Conrad, "Datenhandel und Unternehmenskauf" in *Zeitschrift für Datenschutz* 2016, pp. 1 *et seq*.

36 Cf. H. Zech, "'Industrie 4.0', Rechtsrahmen für eine Datenwirtschaft im digitalen Binnenmarkt" in *Gewerblicher Rechtschutz und Urheberrecht* 2015, p. 1151 (1155).

37 Cf. Zech (Fn. 15) p. 1159. For a criminal law perspective, cf. T. Hoeren, "Dateneigentum, Versuch einer Anwendung von § 303a StGB im Zivilrecht" in *MultiMedia und Recht* 2013, p. 486 *et seq*. For a summary infused with skepticism, see Dorner (Fn. 3).

then fed back into the value-creation process. By way of analogy to the wording of the Federal Constitutional Court's 1983 judgment, according to which there was no longer "unimportant data", one can now state that there is now no more "worthless data".[38]

Proposals for the creation of such a data law range from making a very general analogy to the notion of property in the BGB to concepts based on copyright law.[39] Some are conceived of in the form of a digital register law,[40] but there are also more general proposals and approaches to data law borrowed from the sphere of intellectual property.[41] In 2014, the EU Commission also undertook to examine the need for regulation with regard to individual questions of "data ownership".[42] In addition, there are, of course, also voices which completely reject a new system of data law.[43]

The lines of argument in academic discussion are manifold and tangled: Should such a law also cover personal data? If not, what is the sense in it? How can it be reconciled with the dictum of the Federal Constitutional Court that there can be no extensive right of control, i.e. ownership, over (personal) data in order to prevent social communication?[44] Would it lead to data protection only for the wealthy,[45] or could the resulting commercialisation of data protection to a more efficient structure of data protection? What impact would such a law have on knowledge allocation in a society and would it promote or inhibit innovation? How, in the end, would it affect the law on the ownership of data storage devices, which one was to prevail in a situation of conflict? And so on and so forth.

---

38 Regarding "raw data" cf. Zech (Fn. 15), p. 1156.

39 T. Hoeren, "Dateneigentum, Versuch einer Anwendung von § 303a StGB im Zivilrecht", in *MultiMedia und Recht* 2013, p. 486 et seq.

40 H. Zech, *Informationen als Schutzgegenstand*, Tübingen, 2012, p. 437.

41 See, for example, W. Kerber, "A New (Intellectual) Property Right for Non-Personal Data? An Economic Analysis" in *Gewerblicher Rechtschutz und Urheberrecht Int.* 2016, p. 989.

42 Communication from the Commission to the European Parliament, the Council, the European Economic and Social Committee and the Committee of the Regions, Towards a thriving data-driven-economy, COM(2014) 442, p. 12.

43 See T. Heymann, "Rechte an Daten, Warum Daten keiner eigentumsrechtlichen Logik folgen" in *Computer und Recht* 2016, p. 650 (657).

44 BVerfG *Neue Juristische Wochenschrift* 1984, p. 419 (422).

45 N. Härting, "Acht Thesen zum Dateneigentum" in *Computer und Recht Online*, available at http://www.cr-online.de/blog/2016/02/17/acht-thesen-zum-dateneigentum/ (last accessed 30.11.2016).

The example of an agricultural machine is often used to show the need for "data ownership". The question under discussion is what is the legal basis for the operator of an agricultural machine to “possess“ the data[46] collected by the machine. It soon becomes clear that the traditional legal categories of *ownership* and *possession* and statutory acquisition by means of *connection* and *mixing* of goods are only applicable to data to a very limited extent.[47] That is only partly the case because even today the law is not settled on whether the owner of a data carrier also owns the data stored on that data carrier.[48]

Lastly, with the increasing exchange of data between machines and entire companies – and beyond – within a digitized value chain, questions of data security will definitely become a central concern as well.[49] This is also due to the fact that as non-personal data becomes increasingly important in financial terms, data security, i.e. the technological protection of data and its transmission, will correspondingly become an ever more important measure for the prevention of liability risks.[50]

The discussion of how the law treats data in inter-company relations, and how it treats non-personal data in particular, has in the meantime become a particularly dynamic one. Whether the legislature decides to amend norms on a one by one basis or dares to rework the entire law, cannot yet be foreseen.

---

46 L. Grosskopf, “Rechte an privat erhobenen Geo- und Telemetriedaten“ in: *IP-Rechtsberater* 2011, p. 259 (260.).

47 T. Heymann, “Rechte an Daten, Warum Daten keiner eigentumsrechtlichen Logik folgen“ in *Computer und Recht* 2016, p. 650.

48 Supporting the idea that the owner of the carrier owns the data on it, see BGH *Neue Juristische Wochenschrift* 1993, p. 2436. For the contrary argument, see C. Stresemann, “§ 90“ in K. Johannsen *et al.* (eds.), *Das Bürgerliche Gesetzbuch mit besonderer Berücksichtigung der Rechtsprechung des Reichsgerichts und des Bundesgerichtshof*, Berlin, 12th ed. 1982, paragraph 25.

49 Cf. also. A. Fischer, M. Gröne & Chr. Stüble, “Wie schützt sich die Industrie vor Wirtschaftsspionage 4.0, Innovatives Digital Rights Management – Vertrauenswürdiger Datenaustausch in komplexen Wertschöpfungsnetzwerken der Automobilindustrie“ in *Datenschutz und Datensicherheit* 2015, p. 657 *et seq*. and H. Junker, “IT-Sicherheit für Industrie 4.0 und IoT, Aktuelle Bedrohungslagen und Herausforderungen der Smart Factory“ in *Datenschutz und Datensicherheit* 2015, p. 647 et *seq*.

50 For a treatment of liability risk issues as well as the exchanges of declarations of intent by autonomous systems, see: O. Sosnitza, “Das Internet der Dinge – Herausforderung oder gewohntes Terrain für das Zivilrecht“ in *Computer und Recht* 2016, p. 764.

F. Summary & Outlook

Digitization, autonomous systems and *Industrie 4.0* are not the first highly disruptive phenomena in the history of civilization which cause the law to react and adapt to the far-reaching challenges facing it. A particularly instructive example is the introduction of railways, to which the legislature reacted in 1871 with the introduction of strict liability into the tort liability law (*Reichshaftpflichtgesetz*). Not unlike in that historical episode, the law of today must develop its ability to promote and steer innovation in the face of a new data-centric economy.

The biggest challenge to data law and data protection, against the backdrop of "*Wirtschaft 4.0*", is probably the sheer amount of data that new products and services create, because these products themselves are increasingly based on data. The law is still responding to these developments using traditional methods and will continue to do so for some time. Nevertheless, it is worthwhile to consider whether and how the handling of data in a digitized and networked world and economy could be regulated in a fundamentally new way. In a data-driven economy and society, in the long term, the relationships between companies and other companies and between customers and employees, will no longer be as clearly differentiated as is the case today (or rather: as it was until recently). In some aspects, data have already shifted from being a necessary condition for value-creation processes to being the end result of value-creation. These changes do not only affect the commercial world, in fact, they can only be understood including their social implications.

Informational self-determination (e.g. as a customer) and co-determination rights in the workplace (of employees) are up for negotiation as is the legal protection of data-based businesses from unfair competition. The first step towards preserving elements of our everyday life and economic reality, such as individual informational self-determination and co-determination in employment contexts, will be entering into a broad societal debate on the implications of the ubiquity of data, to which legal science can certainly try to contribute. If anything, this text wanted to show that this debate should not be led in a black-and-white-manner. It neither is business and industry against privacy nor has it to be technological development against privacy.